Geographies of Tourism and Gl

CW01521685

Series Editors
Dieter K. Müller, Department of Geography, Umeå University, Umea, ...
Jarkko Saarinen, Geography Research Unit, University of Oulu, Oulu, Finland
Carolin Funck, Hiroshima University Graduate School of Humanities
and Social Sciences, Higashi-Hiroshima City, Hiroshima, Japan

In a geographical tradition and using an integrated approach, this book series acknowledges the interrelationship of tourism to wider processes within society and environment. This is done at local, regional, national, and global scales demonstrating links between these scales as well as outcomes of global change for individuals, communities, and societies. Local and regional factors will also be considered as mediators of global change in tourism geographies affecting communities and environments. Thus *Geographies of Tourism and Global Change* applies a truly global perspective highlighting development in different parts of the world and acknowledges tourism as a formative cause for societal and environmental change in an increasingly interconnected world.

The scope of the series is broad and preference will be given to crisp and highly impactful work. Authors and Editors of monographs and edited volumes from across the globe are welcome to submit proposals. The series insists on a thorough and scholarly perspective, in addition authors are encouraged to consider practical relevance and matters of subject specific importance. All titles are thoroughly reviewed prior to acceptance and publication, ensuring a respectable and high quality collection of publications.

Jarkko Saarinen • Berendien Lubbe
Naomi N. Moswete

Editors

Southern African Perspectives on Sustainable Tourism Management

Tourism and Changing Localities

 Springer

Editors
Jarkko Saarinen
Geography Research Unit
University of Oulu
Oulu, Finland

Berendien Lubbe
Department Historical and Heritage Studies
University of Pretoria
Pretoria, South Africa

School of Tourism and Hospitality
University of Johannesburg
Johannesburg, South Africa

Naomi N. Moswete
Department of Environmental Science
University of Botswana
Gaborone, Botswana

ISSN 2366-5610   ISSN 2366-5629   (electronic)
Geographies of Tourism and Global Change
ISBN 978-3-030-99437-2   ISBN 978-3-030-99435-8   (eBook)
https://doi.org/10.1007/978-3-030-99435-8

This Springer imprint is published by the registered company Springer Nature Switzerland AG
The registered company address is: Gewerbestrasse 11, 6330 Cham, Switzerland

Preface and Acknowledgements

This edited collection focuses on tourism, sustainability and local change in south-ern Africa. The book aims to offer versatile perspectives that address various changes and challenges in the southern African tourism landscape. The tourism industry is an increasingly important economy in the region, and it is creating a multitude of changes for communities and the environment. The industry itself also faces significant changes and challenges. The key drivers of change that include globalisation, climate change, and deepening global and regional inequalities form the context for the diverse and exciting set of case studies from the region. The book offers a case study–driven approach to sustainability and change management needs in tourism development in local community contexts. The case study chapters are linked through the book's focus on sustainable tourism and local community devel-opment. The book emphasises explicitly and implicitly the need to understand both global change and local contexts in sustainable tourism development.

This book results from the small collaborative project funded by the Southern African – Nordic Centre (SANORD) in 2018–2019. The project 'Tourism for Development? Perspectives to Sustainable Tourism Management in Global South' was coordinated by the University of Pretoria. The partners were: the University of Botswana, the University of Eastern Finland, the University of Johannesburg and the University of Oulu (FIN). The concrete-level aims of the project were to bring together supervisors and graduate students by organising graduate school sympo-siums. The first meeting was held at the Hillcrest Sports Campus of the University of Pretoria in October 2018. The second meeting was a part of the 12th SANORD Annual Scientific Conference hosted by the University of Botswana in Gaborone in September 2019. Furthermore, based on these meetings and created networks, the plan was to process an edited research-based book, hence this book.

For supporting the publishing of this book, we would like to thank Springer and especially Prasad Gurunadham, Evelien Bakker and Bernadette Deelen-Mans for their support and patience. Although this volume does not focus on the COVID-19 issues, as it has been initiated before the pandemic, the crisis has resulted in chal-lenges pertaining to the book process in its schedule and composition. Our thanks also go to the book series editors Caroline Funck and Dieter Muller. As the editors of the book, we would like to acknowledge several individuals and partners in the development of this book.

Jarkko Saarinen would like to thank many colleagues, postgraduate students, friends and institutions as well as other actors from the southern African region who have contributed to the ideas related to sustainability needs in tourism. In particular, he would like to thank Chris and Jayne Rogerson at the University of Johannesburg, Robin Nunkoo at the University of Mauritius, Julius Atlhopheng at the University of Botswana, and C. Michael Hall at the University of Canterbury, New Zealand. The Universities of Johannesburg, Botswana and Pretoria and the colleagues there, Villa Pablo and Allesverloren, have also contributed towards the development of this book. Naomi Moswete would like to thank the Research and Development Unit at the University of Botswana for being part of the organising team towards the SANORD conference in 2019, which contributed immensely towards the ideas of this book project. Many thanks go to all those who took part in the process of the book, such as staff from the Department of Environmental science, namely Masego Mpotokwane, David Lesolle and Ditiro Moalafhi, and a score of graduate students. Special thanks also go to Mary Ellen Kimaro at the University of Namibia. Berendien Lubbe would like to thank SANORD for creating the opportunity for collaboration on this project and book, allowing academics to share ideas and work together on this project. I appreciate the guidance and input of Jarkko Saarinen as our lead editor and Naomi Moswete for her insightful thoughts on the book. I would also like to thank Prof Karen Harris, Head of the Department of Historical and Heritage Studies at the University of Pretoria, for her constant support in these endeavours.

Collectively, we would like to express our appreciation to all who contributed to writing the chapters or peer reviewing them. The overall book has been externally peer reviewed based on the series editors. In addition, each chapter was externally peer reviewed by independent experts in their field.

Kiiminki, Finland   Jarkko Saarinen

Mochudi, Botswana   Naomi N. Moswete

Knysna, South Africa   Berendien Lubbe

Contents

1   Sustainable Tourism Development in the Southern African
    Context: An Introduction.................................... 1
    Jarkko Saarinen, Berendien Lubbe, and Naomi N. Moswete

2   Sustainability Consciousness in the Hospitality Sector
    in Zimbabwe 15
    Ngoni Courage Shereni, Jarkko Saarinen,
    and Christian M. Rogerson

3   In Pursuit of Sustainable Tourism in Botswana:
    Perceptions of Maun Tourism Accommodation Operators
    on Tourism Certification and Eco-Labelling 31
    Godiraone Trompies Motsaathebe and Wame L. Hambira

4   Inbound Tour Operator Participation
    in Sustainable Tourism Practices: A Focus on South Africa........ 47
    Ignatius Ludolph Steyn, Felicite Fairer-Wessels,
    and Anneli Douglas

5   Tourism-Led Inclusive Growth Paradigm:
    Opportunities and Challenges in the Agricultural Food
    Supply Chain in Livingstone, Zambia 61
    Brenda M. K. Nsanzya and Jarkko Saarinen

6   Insourcing the Indigenous Without Outsourcing
    the Story Teller: A Sustainable African Solution 79
    Karen L. Harris and Christoffel R. Botha

7   Assessment of Costs and Benefits of Joint Venture
    Partnerships in Community-Based Tourism Between
    the Private Sector and Goo-Moremi Residents, Botswana......... 91
    Bontle Elijah, Naomi N. Moswete, and Masego A. Mpotokwane

8   Socio-economic Impacts of Community-Based Ecotourism
    on Rural Livelihoods: A Case Study of Khawa Village
    in the Kalahari Region, Botswana............................. 109
    Naomi N. Moswete, Jarkko Saarinen, and Brijesh Thapa

9   Community-Based Tourism as a Pathway Towards Sustainable
    Livelihoods and Well-being in Southern Africa.................. 125
    Alinah Kelo Segobye, Maduo Mpolokang, Ngoni Courage Shereni,
    Stephen Mago, and Malatsi Seleka

10 Changing Environment and the Political Ecology
    of Authenticity in Heritage Tourism: A Case of the Ovahimba
    and the Ju/'Hoansi-San Living Museums in Namibia............. 139
    Isobel Green and Jarkko Saarinen

11 Perspectives on the Applicability of Nexus Thinking
    to Private Protected Areas: A Case Study of Mokolodi
    Nature Reserve, Botswana 153
    James Maradza, Raban Chanda, and Naomi N. Moswete

12 Environmental Change, Wildlife-Based Tourism
    and Sustainability in Chobe National Park, Botswana 169
    Maduo O. Mpolokang, Jeremy S. Perkins, Jarkko Saarinen,
    and Naomi N. Moswete

13 The Impact of Rhino Poaching on the Economic
    Dimension of Sustainable Development in Wildlife Tourism 187
    Berendien Lubbe

14 Locational Heterogeneity in Climate Change Threats
    to Beach Tourism Destinations in South Africa 199
    Jonathan Friedrich, Jannik Stahl, Gijsbert Hoogendoorn,
    and Jennifer M. Fitchett

15 Sustainable Tourism Development Needs
    in the Southern African Context: Concluding Remarks........... 215
    Jarkko Saarinen, Naomi N. Moswete, and Berendien Lubbe

Chapter 1
Sustainable Tourism Development in the Southern African Context: An Introduction

Jarkko Saarinen, Berendien Lubbe, and Naomi N. Moswete

1.1 Introduction

Since the 1990s, sustainability has become a central notion in tourism and related socio-economic development discussions, strategies and policies (Hall & Lew, 1999; Scheyvens, 2011). The idea of sustainability has been firmly incorporated into the tourism industry's critical policies at various planning scales and development settings (Bramwell, 2011; Sharpley, 2000, 2020). Hall (2011, p. 650) stated that the sustainability dimension has been "one of the great success stories of tourism research". This success element reflects in many international tourism policies and planning documents. The United Nations (UN) (2017), for example, has highlighted the broader transformative role that sustainable tourism might play in economic development and inclusive growth in the different subregions of Africa, including southern Africa.

The established connection between tourism and sustainable development has justified the industry's growth needs. However, the idea of sustainability in tourism or the tourism industry's role in sustainable development has been, and still is, debatable and understood in various ways. Indeed, there are different and

J. Saarinen (✉)
Geography Research Unit, University of Oulu, Oulu, Finland

School of Tourism and Hospitality, University of Johannesburg, Johannesburg, South Africa
e-mail: jarkko.saarinen@oulu.fi

B. Lubbe
Department Historical and Heritage Studies, University of Pretoria, Pretoria, South Africa
e-mail: berendien.lubbe@up.ac.za

N. N. Moswete
Department of Environmental Science, University of Botswana, Gaborone, Botswana
e-mail: moatshen@ub.ac.bw

© The Author(s), under exclusive license to Springer Nature
Switzerland AG 2022
J. Saarinen et al. (eds.), *Southern African Perspectives on Sustainable Tourism Management*, Geographies of Tourism and Global Change,
https://doi.org/10.1007/978-3-030-99435-8_1

1

sometimes competing understandings of sustainable tourism and its goals (Saarinen, 2014, 2020). Coccossis (1996), for example, has identified four different ways to interpret and position tourism in sustainable development. These are *sectoral* (the industry's overall perspective), *ecological*, *destination competitiveness* and *strategical*.

Furthermore, according to Clarke (1997, p. 229), sustainable tourism thinking has evolved from being understood as the opposite to conventional mass tourism in a position of convergence in which sustainable tourism is a goal "that all tourism, regardless of scale, must strive to achieve". There are different approaches and understandings of sustainable tourism with little standard agreement on the concept. While new ideas and conceptualisations have emerged over time, the previous ones may still exist and play a role in specific academic and development contexts. For instance, in his conceptualisation and understanding of sustainable tourism, Butler (1993, p. 29) says sustainable tourism refers to tourism that:

> is developed and maintained in an area (community, environment) in such a manner and at such a scale that it remains viable over an infinite period and does not degrade or alter the environment (human and physical) in which it exists to such a degree that it prohibits the successful development and wellbeing of other activities and processes.

This description is a widely used and comprehensive academic definition for sustainable tourism, which also involves a critical component in local and regional development contexts, indicating the simple fact that tourism is a resource user. Thus, the industry impacts on resources it uses. While tourism has many potentially beneficial characteristics for contributing to sustainable development in reality at local and regional scales, it may not automatically be the most sustainable resource user in the long term (Butler, 1999). In the southern African development context, tourism as a resource user is a crucial issue for consideration (Kimaro & Saarinen, 2019; Lenao & Saarinen, 2015; Moswete et al., 2012; Pillay & Rogerson, 2013; Spenceley, 2008). Tourism may compete with traditional livelihoods that have provided community benefits and wellbeing for a long time (see Chiutsi & Saarinen, 2017; Duim van der et al., 2011; Kavita & Saarinen, 2016; Moswete & Thapa, 2018).

Disagreements in sustainable tourism thinking and development are also ideologically driven. Many scholars consider the current hegemonic understanding of sustainable tourism industry-oriented and based on the neoliberal growth agenda that mainly serves the growing needs of global tourism (Hall, 2009, 2019; Hollenhorst et al., 2014; Saarinen, 2021; Sharpley, 2009, 2020). In this respect, Gössling et al. (2020) have indicated that many supranational organisations in tourism policymaking, like the World Tourism Organization (UNWTO), may simply represent tourism growth advocates using sustainability rhetoric (see also Gössling et al., 2016; Scheyvens, 2011). In contrast to this growth-orientated sustainable tourism agenda, Hall et al. (2015, p. x) contend that truly sustainable tourism should be seen as "a subset of sustainable development", and a "tourism system that encourages qualitative development, with a focus on quality of life and wellbeing measures, but not aggregate quantitative growth to the detriment of natural capital".

Due to the many different conceptualisations of sustainable tourism, this edited book has not aimed at having one way to understanding sustainability in tourism; thus, each chapter may have its particular connotation towards sustainability in tourism. However, the book shares a common understanding that the key elements (ecologic, social and economic) and principles (holism, equity and future orientation) of sustainable development should integrate into the tourism industry's operations and specific relations with diverse local communities and environments. Furthermore, there is an (explicit or implicit) emphasis on the policy aims that would highlight the potential role of tourism in the United Nations Sustainable Development Goals (SDGs), adopted in 2015 (United Nations, 2015). These interconnections underline the importance and responsibility of tourism as one of the largest industries in southern Africa, with the potential to contribute to and make a difference to sustainable development in general. In the next section, we will briefly outline the SDGs in the context of tourism, followed by a discussion on future sustainable tourism perspectives in southern Africa. After that, we briefly introduce the chapters of the book.

1.2 Sustainable Development Goals and Tourism Development

The SDGs define the agenda for global development towards 2030 by addressing pertinent issues such as poverty, inequality, global climate change, environmental degradation, and peace and justice (United Nations, 2015). There are 17 goals (Table 1.1) and 169 specific targets. These goals and targets focus on a global scale emphasising that global development challenges depend on the actions taking place in the Global South and North.

The SDGs provide many opportunities for the tourism industry to bring positive change and contribute to long-term sustainability (Rasoolimanesh et al., 2020; Scheyvens, 2018). Undoubtedly, tourism has a high potential to achieve good (or bad) outcomes for destination communities and environments (Hall, 2019; Saarinen & Manwa, 2008; Saarinen, 2019). In order to create and cultivate positive outcomes, Scheyvens (2018, p. 341) has called for tourism scholars "to consider how we might utilise the SDGs to analyse the linkages between tourism and sustainable development in a wide range of contexts and at different scales." At a policy level, this call was supported by the United Nations General Assembly's initiative during the 'International Year of Sustainable Tourism for Development' in 2017 (see UNWTO, 2017) that highlighted the importance of tourism in fostering development and better understanding among people. According to the initiative, tourism could focus on three specific SDGs: Promote sustained, inclusive and sustainable economic growth, full and productive employment and decent work for all (SDG 8); Ensure sustainable consumption and production patterns (SDG 12); and Conserve and sustainably use the oceans, seas and marine resources for sustainable development (SDG 14). These are important goals with which the industry can work. Concerning SDG8, for

Table 1.1 The UN Sustainable Development Goals (SDGs)

Goal 1: No Poverty: Economic growth must be inclusive of providing sustainable jobs and promoting equality.

Goal 2: Zero Hunger: The food and agriculture sectors offer critical development solutions and are central to hunger and poverty eradication.

Goal 3: Good Health and Well-being: Ensuring healthy lives and promoting wellbeing for all ages is essential to sustainable development.

Goal 4: Quality Education: Obtaining a quality education is the foundation to improving people's lives and sustainable development.

Goal 5: Gender Equality: Gender equality is not only a fundamental human right but a necessary foundation for a peaceful, prosperous and sustainable world

Goal 6: Clean Water and Sanitation: Clean, accessible water for all is an essential part of the world we want to live in.

Goal 7: Affordable and Clean Energy: Energy is central to nearly every major challenge and opportunity.

Goal 8: Decent Work and Economic Growth: Sustainable economic growth will require societies to create conditions for quality jobs.

Goal 9: Industry, Innovation and Infrastructure: Investments in infrastructure are crucial to achieving sustainable development.

Goal 10: Reduced Inequality: Policies should be universal in principle to reduce inequalities, paying attention to the needs of disadvantaged and marginalised populations.

Goal 11: Sustainable Cities and Communities: There needs to be a future in which cities provide opportunities for all, with access to essential services, energy, housing, transportation and more.

Goal 12: Responsible Consumption and Production: There needs to be responsible production and consumption.

Goal 13: Climate Action: Climate change is a global challenge that affects everyone everywhere.

Goal 14: Life Below Water: Careful management of this essential global resource is a vital feature of a sustainable future.

Goal 15: Life on Land: Sustainably manage forests, combat desertification, halt and reverse land degradation, halt biodiversity loss.

Goal 16: Peace and Justice Strong Institutions: Access to justice for all and building effective, accountable institutions at all levels.

Goal 17: Partnerships to Achieve the Goal: Revitalise the global partnership for sustainable development.

Source: United Nations, 2015

example, the United Nations Conference and Trade Development (UNCTAD) has indicated that "tourism employment is not gender neutral in Africa, as women and men do not necessarily have the same opportunities in and benefits from the sector" (UNCTAD, 2017, p. 90). While this applies to SDG8, it strongly connects with Goal 5: Gender Equality. Thus, those initial three highlighted SDGs for tourism to work with (UNWTO, 2017) provide a minimal perspective on the potential relationships between tourism and SDGs.

In this respect, the World Bank Group (2017) has expanded the potential connections between tourism and the SDGs. They indicate that tourism can work for sustainable development based on five core pillars: (1) sustainable economic growth; (2) social inclusiveness, employment and poverty reduction; (3) resource efficiency, environmental protection and climate; (4) cultural values, diversity and heritage; and (5) mutual understanding, peace and security. While these pillars widen the potential connections and scope between tourism and the SDGs, they also create possible conflicts between tourism and sustainable development. The sustainable economic growth pillar, for example, is based on a condition that tourism stimulates Gross Domestic Product (GDP) growth, international trade and investments. All these conditions are in line with the growth ideology associated with current neoliberal economic thinking (Daly, 1996; Hall, 2019), creating a potential conflict with climate action (SDG 13), for example (see Saarinen, 2020).

Many international development agencies consider sustainable tourism a good tool for promoting the SDGs and for "benefitting communities in destinations around the world" (World Bank Group, 2017, p. 5). However, many scholars have been more critical (Bianchi, 2018; Gössling et al., 2020; Hall et al., 2015; Mosedale, 2014, 2015; Scheyvens, 2011; Scheyvens & Momsen, 2008; Schilcher, 2007). Indeed, there is a growing field of research on the connections and misconnections between economic growth (including tourism development) and SDGs. For example, Stephen McCloskey (2015, p. 192) highlighted the need to rethink the currently dominant neoliberal development model between the Global North and South to achieve the SDGs by 2030 because of its "illicit financial flows, unfair trade rules, climate change and corporate power."

In a tourism development context, Boluk et al. (2019) have critically debated the potential of the tourism industry to serve the SDGs for more just futures on a destination scale. They highlight the need for: critical tourism scholarship, more profound engagement with indigenous (or local) perspectives; degrowth and the circular economy; better governance and planning, and ethical consumption. For instance, Hall (2019) and Scheyvens et al. (2021) have also been sceptical about the capacity of the SDGs to provide a guiding framework for sustainable development in tourism. For Hall, the industry's 'managerial ecological' approach and the critical policymaking institutions have enabled increasingly hostile environmental and social changes in destinations and the whole tourism system. Hall calls for critical rethinking and restructuring of human-environment relations to change this negative cycle, based on a proactive search for alternative development paths to the current neoliberal agenda. If not, there is no justifiable hope for tourism to be sustainable in the future.

1.3 Sustainable Tourism Development and Management in Southern Africa

Before the wake of COVID-19 pandemic, the tourism industry was a highly relevant and characteristic socio-economic element in most southern African countries. While there have been challenges to creating local well-embedded tourism operations (see Anderson, 2010, 2011; Mbaiwa, 2005; Moswete & Thapa, 2018; Novelli

& Gebhardt, 2007; Rogerson et al., 2013; Saarinen, 2016), the industry also supported communities and local development (Duim van der et al., 2011; Kimaro & Saarinen, 2019; Manwa & Manwa, 2014; Moswete et al., 2012; Moswete et al., 2020; Rogerson, 2006; Rogerson & Saarinen, 2018). The pandemic has caused severe problems for regional tourism industries and local supply chains, including people who work(ed) with tourism and tourism-dependent communities. However, paradoxically, the crisis highlighted the importance of tourists and tour operators that are no longer taken for granted. Instead, their socio-economic values, roles and networks have become well illuminated to all.

It is still too early to have a well-informed view on how (international) tourism will re-start and how the southern African tourism industry will recover from the continuing crisis. Internationally, there are two major views on post-COVID-19 tourism development. The first one emphasises the urgency of global tourism to return to its pre-COVID-19 growth path, advocated by international agencies such as UNWTO (see Gössling et al., 2020). The second outlook focuses on how we should try to make post-COVID-19 tourism development more sustainable than it was (see Brouder, 2020; Rogerson & Baum, 2020; Sigala, 2020). Prideaux et al. (2020, p. 668) have highlighted that for the tourism industry's future we should "look beyond the temptation of adopting strategies based on a return to the normal of the past". Instead, Prideaux et al. suggest that we should seek to develop the tourism industry to respond and contribute positively to the transformative needs of the global economy.

It is quite probable that the temptation to generate and support fast growth and a return to the previous growth path is too high for governments, politicians and regional policymakers. Perceptibly, the industry is more than keen to get back to business-as-usual. Still, it is just as likely that many more tourism scholars (and policymakers) have realised the need for sustainability and resilience thinking and the value of good governance and management in tourism planning and development. Rogerson and Baum (2020, p. 733) have observed that "the most important message is imperative for a post-COVID 19 African tourism to become more aligned with SDGs." They also strongly support Ezeh and Fonn's (2020) call for better interaction and collaboration between governments, policymakers, and the African research community in creating future tourism in Africa. In this respect, Rogerson and Baum's proposed research agenda includes critical issues in sustainable tourism management in the southern African perspective, such as the need to support regional and domestic tourism, community-based tourism development and creating resilience in the informal sector and climate change contexts. The climate change process represents a slow-onset disaster or crisis affecting the southern African region at a different pace in time and space, resulting in potentially severe impacts and challenges for tourism development in different parts of the region (Hoogendoorn & Fitchett, 2018; Saarinen et al., 2020). Similarly, Senbeto et al. (2021) call attention to research and practical gaps associated with issues such as abject poverty alleviation and the Sustainable Development Goals (SDGs) and crisis and crisis management in African tourism and hospitality studies in future.

The call for a more sustainable and diverse tourism industry is not new in the southern African context (Moswete et al., 2012; Rogerson & Rogerson, 2010; Rogerson & Visser, 2004, 2011, 2020; Saarinen & Manwa, 2008). Furthermore, Saarinen et al. (2009) recommended the development of sustainable tourism education in southern African tertiary education institutions. Thus, this book aims to integrate sustainable tourism research and education perspectives by drawing attention to sustainability needs in southern African tourism planning, development and management. For the most part, the focus of this book is on tourism, sustainability and local change to provide new perspectives addressing change and change management in the southern African tourism landscape. This research-based edited collection of case studies cannot cover the overall region of southern Africa and focus mainly on the 'southern' countries of the region: Botswana, Namibia, South Africa, Zambia and Zimbabwe.

While the tourism industry is an increasingly important economy creating multiple changes for communities and environments in the region (Rogerson & Rogerson, 2010), the industry itself faces many changes and challenges. In addition to the COVID-19 crisis, the key drivers of change include globalisation, climate change, and deepening global and regional inequalities, which form the context for the region's diverse and exciting set of case studies. The book offers a case study driven approach to sustainability and change management needs in tourism development in community contexts. Overall, the book emphasises the need to understand global change and local contexts in sustainable tourism development and management.

1.4   The Contributions

The book does not have separate sections for different approaches or interpretations of sustainable development and sustainability needs in tourism. Instead, the perspective on sustainability aspects in tourism transforms through the critical elements of sustainable development discussions, namely economic, socio-cultural and environmental perspectives. The book begins with a focus on sustainable tourism development based on the industry's and operators' perspectives, followed by a discussion on community issues and participation in tourism. Finally, environmental aspects and changes in the southern African tourism scene are covered. The chapters are research-based and context-driven case studies involving different views on sustainable tourism development and management. The environmental aspects of sustainable tourism are also partly discussed and illuminated through operators or tourists' perceptions and preferences. However, the overall emphasis of the work moves from economic issues to socio-cultural ones and, finally, environmental and natural resource-based views on sustainable tourism development and management in southern Africa.

After this introductory chapter, Shereni, Saarinen and Rogerson analyse sustainability consciousness with a focus on the hospitality sector in Zimbabwe. They aim to understand how stakeholders in the hospitality sector perceive sustainability and related practices in hotels, lodges and guest houses. The key findings indicate that operators in the hospitality sector recognise that they use finite resources, and the management level staff are aware of sustainability needs in tourism development. However, the results also indicate that sustainability awareness among employees is generally low. Godiraone Motsaathebe and Wame Hambira continue the analysis of the industry views on sustainability by studying the perceptions of tourism accommodation operators regarding certification and eco-labelling in tourism in Maun, Botswana. The eco-labelling based on the Botswana Ecotourism Certification System is viewed positively by the operators willing to participate in the system. However, there is a severe lack of awareness about the certification system.

Similarly, in their chapter, Shereni, Saarinen, and Rogerson recommend that policymakers intensify awareness-raising and devise strategies to incentivise companies to take the necessary action towards a sustainability path in tourism development. Ignatius Steyn, Felicite Fairer-Wessels and Anneli Douglas move the tourism operator views to the South African context. They state that inbound tour operators can play a crucial role in sustainable tourism development. The operators provide the link between the supply and demand of tourism products and services. Their case study identifies some of the critical gaps in sustainability and outlines potential strategies to overcome these gaps.

The chapter by Brenda Nsanzya and Jarkko Saarinen focuses on a tourism-led inclusive growth paradigm by analysing opportunities and challenges in the agricultural food supply chain in Livingstone, Zambia. By doing so, they identify facilitators and barriers to sustainable market linkages between tourism and local agriculture. The results demonstrate that tourism-agriculture market linkages exist, but they are weak and fragmented, resulting in low positive, inclusive growth outcomes in the agricultural sector. Karen Harris and CR Botha further connect tourism operator and community views by discussing the transformation of heritage tourism by including the 'Indigenous Story Teller' (IST) within the ambit of the regulated tourist guiding sector. They use a case example from the Northern Cape province in South Africa by suggesting incorporating community voices into the tourist experience.

Furthermore, Harris and Botha argue that a mutually beneficial relationship is needed for the effective involvement of ISTs in the broader heritage tourism realm. In their chapter, Bontle Elijah, Naomi Moswete and Masego Mpotokwane discuss the costs and benefits of joint venture partnerships in community-based tourism. Their case study is from the Goo-Moremi community, Botswana. The authors employ social exchange and empowerment theories to understand the effects of joint venture partnerships in community-based tourism. The key findings indicate that community members have received limited benefits from heritage tourism and the joint partnerships in managing the local heritage site. Some challenges also exist, such as uneven power relations between the community and the operator, resulting from, for example, a lack of commitment by the local community members.

Naomi Moswete, Jarkko Saarinen and Brijesh Thapa focus on socio-economic impacts of community-based ecotourism on rural livelihoods with a case study of

Khawa village in the Kalahari region, Botswana. Community-based tourism has become an increasingly important activity in Botswana, but its socio-economic benefits for local development have often been questioned. The results from this remote Khawa village indicate that some benefits exist, such as seasonal job opportunities, but that there are also negative impacts from tourism that should be appropriately managed in the foreseeable future. Alinah Segobye, Maduo Mpolokang, Ngoni Shereni, Stephen Mago and Malatsi Seleka turn the community-based tourism discussion to a general Southern African Development Community (SADC) level by using two case examples of the Mababe concession area in the Ngamiland District, Botswana, and Hwange National Park, Zimbabwe. Specifically, they aim to explore how initiatives like trans-border frontier parks and community-based natural resources-based management (CBNRM) programmes could promote more inclusive local development. Furthermore, the chapter highlights the need to incorporate conflict management and peacebuilding into biodiversity conservation through CBNRM programmes.

Isobel Green and Jarkko Saarinen continue discussing community aspects in tourism development and management. They analyse the role and impacts of a changing environment on cultural performances and authenticity in heritage tourism with two empirical case examples from Namibia: the Ovahimba and the Ju/'Hoansi-San Living Museums. The chapter utilises a political ecology perspective to understand the entwined nature of local culture. They discuss how heritage elements are produced and displayed in these living museums in Namibia. Furthermore, the chapter analyses how displayed heritage tourism and its produced authenticity have been affected by environmental changes at the case study sites. James Maradza, Raban Chanda and Naomi Moswete retain and deepen the environmental and natural resource views in community-tourism relations by studying the applicability of so-called Nexus Thinking. Their case is Mokolodi Nature Reserve, Botswana, a private protected area surrounded by local communities. They indicate that mutually beneficial linkages exist between the Reserve and the community, including empowerment of communities in development and planning processes, although community members expected more concrete collaborative processes.

Maduo Mpolokang, Jeremy Perkins, Jarkko Saarinen and Naomi Moswete shift the focus to environmental change for wildlife-based tourism and sustainability in the Chobe National Park, Botswana and the impact of such change. They emphasise that in the context of wildlife-based tourism development and natural resources, creating knowledge on the nature and scale of environmental change is fundamental to sustainable tourism and related responsive policy formulations and planning measures. Berendien Lubbe continues the sustainable wildlife tourism discussion by analysing the economic impact of rhino poaching in South Africa, specifically in the Kruger National Park. Based on the results, the economic impact of rhino poaching on tourism and tourists highlights three main issues, which include: longer-term loss of tourism revenue through a decrease in arrivals which impacts both conservation efforts and socio-economic development in communities; the short-term adverse effects on the tourists' experience; and the loss of the inherent value of the animal shown to be far more challenging to quantify.

Finally, Jonathan Friedrich, Jannik Stahl, Gijsbert Hoogendoorn and Jennifer Fitchett advocate the importance of the environment and its condition for sustainable tourism development and management in southern Africa. Specifically, they highlight the industry's dependency on the climate and day-to-day weather, making the tourism sector vulnerable to and threatened by the estimated impacts of global climate change. In this context, they provide an overview of heterogeneity in both climate change threats and beach tourists' perceptions of climate and weather at nine destinations along the South African coastline. A tourist survey predicted that extreme climate events and diseases are the primary determinants that visitors indicated would prompt them to cancel their trips. Thus, the identified threats require coastal destinations to develop local and dynamic adaption strategies to cope with climate change and sustain tourism's economic contribution in the southern African region.

The last chapter by the editors concludes the key aspects and research needs for sustainable tourism development and management in southern Africa. It is noted that while sustainable tourism has become an established field of research in the region with numerous supporting development policies and strategies, the relationships between the tourism industry and localities have remained complex and even controversial. Despite the challenges, however, the past research and the cases of this book demonstrate that it is possible to create positive synergies between tourism and localities in sustainable development. The chapter concludes that a better implementation of tourism development to serve the SDGs is critical.

References

Anderson, W. (2010). Determinants of enclave travel expenditure. *Tourism Review, 65*(3), 4–15. https://doi.org/10.1108/16605371011083495

Anderson, W. (2011). Enclave tourism and its socio-economic impact in emerging destinations. *Anatolia, 22*(3), 361–377. https://doi.org/10.1080/13032917.2011.633041

Bianchi, R. (2018). The political economy of tourism development: A critical review. *Annals of Tourism Research, 70*, 88–102. https://doi.org/10.1016/j.annals.2017.08.005

Boluk, K. A., Cavaliere, C. T., & Higgins-Desbiolles, F. (2019). A critical framework for interrogating the United Nations Sustainable Development Goals 2030 Agenda in tourism. *Journal of Sustainable Tourism, 27*(7), 847–864. https://doi.org/10.1080/09669582.2019.1619748

Bramwell, B. (2011). Governance, the state and sustainable tourism: A political economy approach. *Journal of Sustainable Tourism, 19*(4–5), 459–477. https://doi.org/10.1080/0966958 2.2011.576765

Brouder, P. (2020). Reset Redux: Possible evolutionary pathways towards the transformation of tourism in a COVID-19 world. *Tourism Geographies, 22*(3), 484–490.

Butler, R. W. (1993). Tourism - an evolutionary perspective. In J. G. Nelson, R. W. Butler, & G. Wall (Eds.), *Tourism and sustainable development: Monitoring, planning, managing* (pp. 27–44). University of Waterloo.

Butler, R. (1999). Sustainable tourism: A state-of-the-art review. *Tourism Geographies, 1*(1), 7–25.

Chiutsi, S., & Saarinen, J. (2017). Local participation in transfrontier tourism: Case of Sengwe community in Great Limpopo Transfrontier Conservation Area, Zimbabwe. *Development Southern Africa, 34*(3), 260–275. https://doi.org/10.1080/0376835X.2016.1259987

Clarke, J. (1997). A framework of approaches to sustainable tourism. *Journal of Sustainable Tourism, 5*, 224–233. https://doi.org/10.1080/09669589708667287

Coccossis, H. (1996). Tourism and sustainability: Perspectives and implications. In G. K. Priestley, J. A. Edwards, & H. Coccossis (Eds.), *Sustainable tourism? European experiences* (pp. 1–21). CAB International.

Daly, H. (1996). *Beyond growth: The economics of sustainable development*. Beacon Press.

Duim van der, R., Meyer, D., Saarinen, J., & Zellmer, K. (Eds.). (2011). *New alliances for tourism, conservation and development in eastern and southern Africa*. Eburon.

Gössling, S., Ring, A., Dwyer, L., Andersson, A. C., & Hall, C. M. (2016). Optimising or maximising growth? A challenge for sustainable tourism. *Journal of Sustainable Tourism, 24*(4), 527–548. https://doi.org/10.1080/09669582.2015.1085869

Gössling, S., Scott, D., & Hall, C. M. (2020). Pandemics, tourism and global change: A rapid assessment of COVID-19. *Journal of Sustainable Tourism, 29*(1), 1–20. https://doi.org/10.1080/09669582.2020.1758708

Ezeh, A., & Fonn, S. (2020, May 27). Sub-Saharan Africa needs to plug local knowledge gap to up its anti-COVID-19 game. *The Conversation*. Retrieved from: https://theconversation.com/sub-saharan-africa-needs-to-plug-local-knowledge-gap-to-up-its-anti-covid-19-game-138917

Hall, C. M. (2009). Degrowing tourism: Décroissance, sustainable consumption and steady-state tourism. *Anatolia, 20*(1), 46–61. https://doi.org/10.1080/13032917.2009.10518894

Hall, C. M. (2011). Policy learning and policy failure in sustainable tourism governance: From first- and second-order to third-order change? *Journal of Sustainable Tourism, 19*(4–5), 649–671. https://doi.org/10.1080/09669582.2011.555555

Hall, C. M. (2019). Constructing sustainable tourism development: The 2030 agenda and the managerial ecology of sustainable tourism. *Journal of Sustainable Tourism, 27*(7), 1044–1060. https://doi.org/10.1080/09669582.2018.1560456

Hall, C. M., & Lew, A. A. (Eds.). (1999). *Sustainable tourism: A geographical perspective*. Longman.

Hall, C. M., Gössling, S., & Scott, D. (Eds.). (2015). *The Routledge handbook of tourism and sustainability*. Routledge.

Hollenhorst, S. J., Houge-MacKenzie, S., & Ostergren, D. M. (2014). The trouble with tourism. *Tourism Recreation Research, 39*(3), 305–319. https://doi.org/10.1080/02508281.2014.11087003

Hoogendoorn, G., & Fitchett, J. M. (2018). Tourism and climate change: A review of threats and adaptation strategies for Africa. *Current Issues in Tourism, 21*(7), 742–759. https://doi.org/10.1080/13683500.2016.1188893

Kavita, E., & Saarinen, J. (2016). Tourism and rural community development in Namibia: Policy issues review. *Fennia, 194*(1), 79–88. https://doi.org/10.11143/4633

Kimaro, E., & Saarinen, J. (2019). Tourism and poverty alleviation in the Global South: Emerging corporate social responsibility in the Namibian nature-based tourism industry. In M. Stone, M. Lenao, & N. Moswete (Eds.), *Natural resources, tourism and community livelihoods in southern Africa* (pp. 123–142). Routledge.

Lenao, M., & Saarinen, J. (2015). Integrated rural tourism as a tool for community tourism development: Exploring culture and heritage projects in the north East District of Botswana. *South African Geographical Journal, 97*(2), 203–216. https://doi.org/10.1080/03736245.2015.1028985

Manwa, H., & Manwa, F. (2014). Poverty alleviation through pro-poor tourism: The role of Botswana forest reserves. *Sustainability, 6*(9), 5697–5713. https://doi.org/10.3390/su6095697

Mbaiwa, J. (2005). Enclave tourism and its socio-economic impact in the Okavango Delta, Botswana. *Tourism Management, 26*, 157–172. https://doi.org/10.1016/j.tourman.2003.11.005

McCloskey, S. (2015). Viewpoint. From MDGs to SDGs: We need a critical awakening to succeed. *Policy and Practice, 20*, 186–194.

Mosedale, J. (2014). Political economy of tourism: Regulation theory, institutions, and governance networks. In A. A. Lew, C. M. Hall, & A. M. Williams (Eds.), *The Wiley Blackwell companion to tourism* (pp. 55–65). John Wiley.

Mosedale, J. (Ed.). (2015). *Neoliberalism and the political economy of tourism*. Routledge.

Moswete, N., Thapa, B., & Darley, W. K. (2020). Local communities' attitudes and support towards the Kgalagadi Transfrontier Park in Southwest Botswana. *Sustainability, 12*, 1524.

Moswete, N., & Thapa, B. (2018). Local communities, CBOs/trusts, and people–park relationships: A case study of the Kgalagadi Transfrontier Park, Botswana. *The George Wright Forum, 35*(1), 96–108. https://www.jstor.org/stable/26452995

Moswete, N., Thapa, B., & Child, B. (2012). Attitudes and opinions of local and national public sector stakeholders towards Kgalagadi Transfrontier Park, Botswana. *International Journal of Sustainable Development & World Ecology, 19*(1), 67–80. https://doi.org/10.1080/1350450 9.2011.592551

Novelli, M., & Gebhardt, K. (2007). Community based tourism in Namibia: 'Reality show' or 'window dressing'? *Current Issues in Tourism, 10*(5), 443–479. https://doi.org/10.2167/ cit332.0

Pillay, M., & Rogerson, C. M. (2013). Agriculture-tourism linkages and pro-poor impacts: The accommodation sector of urban coastal KwaZulu-Natal, South Africa. *Applied Geography, 36*, 49–58. https://doi.org/10.1016/j.apgeog.2012.06.005

Prideaux, B., Thompson, M., & Pabel, A. (2020). Lessons from COVID-19 can prepare global tourism for the economic transformation needed to combat climate change. *Tourism Geographies, 22*(3), 667–678. https://doi.org/10.1080/14616688.2020.1762117

Rasoolimanesh, S. M., Ramakrishna, S., Hall, C. M., Esfandiar, K., & Seyfi, S. (2020). A systematic scoping review of sustainable tourism indicators in relation to the sustainable development goals. *Journal of Sustainable Tourism*. https://doi.org/10.1080/09669582.2020.1775621

Rogerson, C. M. (2006). Pro-Poor local economic development in South Africa: The role of pro-poor tourism. *Local Environment, 11*(1), 37–60. https://doi.org/10.1080/13549830500396149

Rogerson, C. M., & Baum, T. (2020). COVID-19 and African tourism research agendas. *Development Southern Africa, 37*(5), 727–741. https://doi.org/10.108 0/0376835X.2020.1818551

Rogerson, C. M., Hunt, H., & Rogerson, J. M. (2013). Safari lodges and local economic linkages in South Africa. *Africanus Journal of Development Studies, 43*(1), 3–17. https://hdl.handle. net/10520/EJC137493

Rogerson, C. M., & Rogerson, J. M. (2010). Local economic development in Africa: Global context and research directions. *Development Southern Africa, 27*(4), 465–480. https://doi.org/1 0.1080/0376835X.2010.508577

Rogerson, C. M., & Saarinen, J. (2018). Tourism for poverty alleviation: Issues and debates in the Global South. In C. Cooper, S. Volo, W. C. Gartner, & N. Scott (Eds.), *The SAGE handbook of tourism management: Applications of theories and concepts to tourism* (pp. 22–37). SAGE Publications Ltd.

Rogerson, C. M., & Visser, G. (2004). *Tourism and development issues in contemporary South Africa*. Africa Institute of South Africa.

Rogerson, C. M., & Visser, G. (2011). African tourism geographies: Existing paths and new directions. *Tijdschrift voor Economische en Sociale Geografie, 102*(3), 251–259.

Rogerson, J. M., & Visser, G. (2020). Research trends in South African tourism geographies. In J. M. Rogerson & G. Visser (Eds.), *New directions in South African tourism geographies* (pp. 1–14). Springer.

Saarinen, J. (2014). Critical sustainability: Setting the limits to growth and responsibility in tourism. *Sustainability, 6*, 1–17. https://doi.org/10.3390/su6010001

Saarinen, J. (2016). Political ecologies and economies of tourism development in Kaokoland, North-West Namibia. In M. Mostafanezhad, A. Carr, & R. Norum (Eds.), *Political ecology of tourism: Communities, power and the environment* (pp. 213–230). Routledge.

Saarinen, J. (2019). Communities and sustainable tourism development: Community impacts and local benefits creation tourism. In S. F. McCool & K. Bosak (Eds.), *A research agenda for sustainable tourism* (pp. 206–222). Edward Elgar Publishing.

Saarinen, J. (2020). Tourism and sustainable development goals: Research on sustainable tourism geographies. In J. Saarinen (Ed.), *Tourism and sustainable development goals: Research on sustainable tourism geographies* (pp. 1–10). Routledge.

Saarinen, J. (2021). Is being responsible sustainable in tourism? Connections and critical differences. *Sustainability, 13*, 6599. https://doi.org/10.3390/su13126599

Saarinen, J., Becker, F., Manwa, H., & Wilson, D. (2009). Introduction: Call for sustainability. In J. Saarinen, F. Becker, H. Manwa, & D. Wilson (Eds.), *Sustainable tourism in southern Africa: Local communities and natural resources in transition* (pp. 3–190). Channel View.

Saarinen, J., & Manwa, H. (2008). Tourism as a socio-cultural encounter: Host-guest relations in tourism development in Botswana. *Botswana Notes and Records, 39*, 43–53. https://www.jstor.org/stable/41236632

Saarinen, J., Moswete, N., Atlhopheng, J., & Hambira, W. (2020). Changing socio-ecologies of Kalahari: Local perceptions towards environmental change and tourism in Kgalagadi, Botswana. *Development Southern Africa, 37*(5), 855–870. https://doi.org/10.1080/0376835X.2020.1809997

Schilcher, D. (2007). Growth versus equity: The continuum of pro-poor tourism and neoliberal governance. *Current Issues in Tourism, 10*(2–3), 166–193. https://doi.org/10.2167/cit304.0

Scheyvens, R. (2011). *Tourism and poverty*. Routledge.

Scheyvens, R. (2018). Linking tourism to the sustainable development goals: A geographical perspective. *Tourism Geographies, 20*(2), 341–342. https://doi.org/10.1080/14616688.2018.1434818

Scheyvens, R., Carr, A., Movono, A., Hughes, E., Higgins-Desbiolles, F., & Mika, J. P. (2021). Indigenous tourism and the sustainable development goals. *Annals of Tourism Research, 90*, 103260. https://doi.org/10.1016/j.annals.2021.103260

Scheyvens, R., & Momsen, J. H. (2008). Tourism and poverty reduction: Issues for small island states. *Tourism Geographies, 10*(1), 22–41. https://doi.org/10.1080/14616680701825115

Senbeto, D. L., Köseoglu, M. A., & King, B. (2021). Hospitality and tourism scholarship in Africa: A literature-based agenda for future research. *Journal of Hospitality & Tourism Research*. https://doi.org/10.1177/10963480211011540

Sharpley, R. (2000). Tourism and sustainable development: Exploring the theoretical divide. *Journal of Sustainable Tourism, 8*(1), 1–19. https://doi.org/10.1080/09669580008667346

Sharpley, R. (2009). *Tourism development and the environment: Beyond sustainability?* Earthscan.

Sharpley, R. (2020). Tourism, sustainable development and the theoretical divide: 20 years on. *Journal of Sustainable Tourism*. https://doi.org/10.1080/09669582.2020.1779732

Sigala, M. (2020). Tourism and COVID-19: Impacts and implications for advancing and resetting industry and research. *Journal of Business Research, 117*, 312–321. https://doi.org/10.1016/j.jbusres.2020.06.015

Spenceley, A. (2008). Requirements for sustainable nature-based tourism in transfrontier conservation areas: A southern African Delphi consultation. *Tourism Geographies, 10*(3), 285–311. https://doi.org/10.1080/14616680802236295

UN (United Nations). (2015). *Transforming our world: The 2030 agenda for sustainable development. Resolution adopted by the General Assembly on 25 September 2015*. United Nations.

UN (United Nations). (2017). *Economic development in Africa report 2017: Tourism for transformative and inclusive growth*. Retrieved from: http://unctad.org/en/PublicationsLibrary/aldcafrica2017_en.pdf

UNCTAD (United Nations Conference and Trade Development). (2017). *The economic development in Africa report 2017: Tourism for transformative and inclusive growth*. United Nations.

UNWTO (World Tourism Organization). (2017). *Tourism and the sustainable development goals – Journey to 2030, highlights*. UNWTO.

World Bank Group. (2017). *Tourism for development: 20 reasons sustainable tourism counts for development*. The World Bank.

Jarkko Saarinen is a Professor of Human Geography (Tourism Studies) at the University of Oulu, Finland, and Distinguished Visiting Professor (Sustainability Management) at the University of Johannesburg, South Africa, and Extraordinary Professor at the Tourism Management Division, Department of Marketing Management, University of Pretoria. His research interests include sustainable development, sustainable tourism, tourism-community relations and nature conservation studies.

Berendien Lubbe holds PhD in Communication Management and is an Emeritus Professor and Research Associate in the Department of Historical and Heritage Studies at the University of Pretoria in South Africa. Her research currently focuses on contemporary issues in tourism. She has published in numerous journals, and her books on Tourism Distribution and Tourism Management in South Africa have been widely prescribed.

Naomi N. Moswete is a Senior Lecturer in the Department of Environmental Science, University of Botswana. Her research interests include Human geography, tourism as a strategy for rural development, community-based tourism; Transboundary conservation areas –ecotourism nexus, parks–people relationships, heritage management & cultural tourism.

Chapter 2
Sustainability Consciousness in the Hospitality Sector in Zimbabwe

Ngoni Courage Shereni, Jarkko Saarinen, and Christian M. Rogerson

2.1 Introduction

Globally, since the early 1990s the idea of sustainability has been incorporated into the hospitality industry's development and management thinking. There are numerous policies and planning frames that emphasise the role and need for sustainable development in the hospitality industry, specifically in the hotel, resort and event management sectors (see Hobson & Essex, 2001; Ricaurte, 2011; Russo & Fouts, 1997). While the industry's development policies have bought into the principles of sustainable development, sustainable hospitality practices depend on how managers and staff have internalised the core notions and demands of sustainable development in their work (see Dodds & Kuehnel, 2008). Thus, there is a need for a consciousness of the principles and requirements of sustainability in the hospitality sector. The current emphasis on the Sustainable Development Goals (SDGs) in tourism further highlights the need for sustainability consciousness (Saarinen, 2020; Scheyvens, 2018).

N. C. Shereni (✉)
Department of Accounting and Finance, Lupane State University, Lupane, Zimbabwe

School of Tourism and Hospitality, University of Johannesburg, Johannesburg, South Africa

J. Saarinen
Geography Research Unit, University of Oulu, Oulu, Finland

School of Tourism and Hospitality, University of Johannesburg, Johannesburg, South Africa
e-mail: jarkko.saarinen@oulu.fi

C. M. Rogerson
School of Tourism and Hospitality, University of Johannesburg, Johannesburg, South Africa
e-mail: chrismr@uj.ac.za

© The Author(s), under exclusive license to Springer Nature
Switzerland AG 2022
J. Saarinen et al. (eds.), *Southern African Perspectives on Sustainable Tourism Management*, Geographies of Tourism and Global Change,
https://doi.org/10.1007/978-3-030-99435-8_2

Bennet and Bennet (2008) have defined consciousness as heightened sensitivity to, awareness of, and connection with our unconscious mind. Clearly, consciousness and awareness are synonymous terms that are often used interchangeably. Thus, in explaining the relationship between consciousness and awareness, Low (2005) has noted that these two concepts evolved from each other. Therefore, one cannot talk of awareness without looking at consciousness the same way that one cannot discuss consciousness without delving into awareness. Musavengane (2019) has argued that consciousness is linked to perceptions, attitudes, attribution and conception. In this regard, transposing the concept of consciousness to sustainability entails looking at the level of awareness, attitudes, perceptions, and the understanding of sustainability issues.

The World Commission on Environment and Development report provides the most widely used definition of sustainability which reads as "development that meets the needs of the present without compromising the ability of future generations to meet their own needs" (WCED, 1987, p. 43). In the tourism industry, the term sustainable tourism is commonly used to refer to "tourism that takes full account of its current and future economic, social and environmental impacts, addressing the needs of visitors, the industry, the environment and host communities" (UNWTO, 2005, p. 12). Sustainability in the tourism industry is focused on the three pillars, which are environmental, social and economic sustainability, also known as the triple bottom line (Elkington, 1998; Saarinen, 2018, 2019). This need for wider focus also applies for the hospitality industry, as there should be an understanding that the company's responsibility goes beyond profit-making to social justice and ecologically sustainable practices. In this respect, this chapter aims to discuss the awareness of the three pillars of sustainability by hospitality operators in Zimbabwe. In addition, the aim is to identify common sustainable practices in the hospitality sector in Zimbabwe. Within tourism scholarship, the chapter contributes to the limited extant literature on sustainability consciousness in the particular context of southern Africa (see Hambira et al., 2013; Rogerson & Rogerson, 2011; Rogerson & Visser, 2020; Saarinen et al., 2012).

2.2   Sustainability Consciousness

Among the different elements of sustainability thinking, environmental consciousness is probably the most studied and discussed aspect in the hospitality industry (Claver-Cortés et al., 2007; Lee et al., 2010; Radwan et al., 2010; Millar et al., 2012; Minoli et al., 2015; De Carvalho et al., 2015; Prud'homme & Raymond, 2016). In general, the hospitality industry has been seen as creating negative impacts to the environment. A number of studies have been dedicated to determining the hospitality sector's awareness of the impacts it poses to the environment (Hobson & Essex, 2001; Middleton & Hawkins, 1998). Environmental consciousness research looks at the realisation by companies that they ought to adopt environmentally friendly practices to appeal to ecologically conscious customers (Ayuso, 2007; Huang et al.,

2014) and to protect the environment. The hospitality industry is a significant consumer of water and energy and produces a lot of waste, causing environmental damage. Environmental consciousness in this study encompasses water conservation, energy-saving, and proper waste management practices. Some of the specific environmentally friendly practices adopted by environmental conscious operators include waste recycling, waste separation, food waste management, linen reuse policy, low flow showerheads in guest rooms, energy management systems and use of LED lights, among others. Guests are now more conscious of environmental issues than before, incentivising hospitality operators to adopt various sustainable practices (Weaver et al., 2013). Environmental sustainability in the hospitality industry is closely associated with green issues hence the adoption of terms such as green tourism, green hospitality and eco-lodging.

Corporate profit is the primary focus of the economic pillar of sustainability (see Coles et al., 2013). The Nobel laureate Milton Friedman (1970) famously argued that "there is one and only one social responsibility of business - to use its resources and engage in activities designed to increase its profits". Indeed, businesses may exist to make profits, but as noted by the World Business Council for Sustainable Development (2002), businesses cannot succeed if the surrounding communities and societies fail. Furthermore, Davis (1960) has noted that the business is and must remain fundamentally an economic institution, but it also has responsibilities to help society achieve its primary goals. Therefore, the profit-making capacity of a business enables it to exist into the future and be able to carry out its social and environmental responsibility roles in society (Franzoni & Avellino, 2019).

Social sustainability is reflected in the role of the business in society and it encompasses the relationship between the business and society (Kimaro & Saarinen, 2020). Beusch and Björnefors (2014) posited that social sustainability is the less developed pillar of sustainability. Issues like Corporate Social Responsibility (CSR), employment of locals, philanthropy, respect of social values and norms, and making a positive impact on society come to mind when one looks at the social sustainability consciousness of a business. Social sustainability is premised on the idea that businesses operate in the context of certain communities, share resources with them and in some instances impact negatively to these communities (Chiutsi & Saarinen, 2019). Hence, there is a need for the adoption of initiatives that ensure that community members benefit from the businesses operating within their localities.

As there is heightened awareness of sustainability issues among global citizens, business operations of hospitality enterprises have come under increasing scrutiny. Arguably, contemporary businesses should not focus only on economic profits but also on how their operations impact the environment and society. Due to its growth-oriented nature, the hospitality industry generates many undesirable impacts such as emission of greenhouse gases, disposal of waste into the environment, overconsumption of natural resources, and exclusion of the local communities and local businesses into the tourism value chain. Therefore, hospitality operators' awareness and consciousness of the three pillars of sustainability is an important factor in minimising the undesirable impacts caused by the hospitality industry. In this respect, it is crucial for stakeholders in the hospitality industry to understand sustainability consciousness as awareness informs behavioural intentions.

Additionally, awareness of sustainability issues by the hospitality sector drives decisions about the adoption of sustainable practices. In the context of SDGs, every sector of the economy is expected to contribute to the global goals. It is essential that sustainable conscious hospitality operators have the capacity to implement practices serving the SDGs. Bianchi (2020, p. 81) observes that prior to the outbreak of the COVID-19 pandemic "global sustainable tourism dialogues had begun to pivot increasingly around the UNWTO's 2015–2030 sustainable development agenda – framed by the 17 United Nations Sustainable Development Goals – the central premise of which is that the transition to inclusive and sustainable tourism can be engineered through the managed growth of tourism".

2.3   Materials and Methods

This study adopted a mixed-methods approach to understand sustainability consciousness in the hospitality sector in Zimbabwe. Probability proportional to size sampling was done to registered hotels, lodges and guest houses in six purposively selected tourism development zones in Zimbabwe. The six tourism development zones were Harare, Bulawayo, Victoria Falls, Masvingo, Kariba and Manicaland. The sampling procedure generated 125 respondents distributed as follows; 21 hotels, 55 guest houses and 49 lodges. A survey was processed by using a self-administered questionnaire to determine sustainability awareness by management employees in the selected hospitality establishments. The questionnaire contained items measuring awareness of sustainability issues on aspects like employee awareness, guest awareness and organisational awareness on a five-point Likert scale. Descriptive statistics were used to analyse survey data showing the mean scores of each questionnaire item. The demographic characteristics of the respondents were 52% male and 48% females, the majority of respondents were educated up to the level of diploma (36%) and bachelor's degree (34%), most respondents were in the age group 31–40 years (62%), and the largest group of the respondents was occupying supervisory roles (37%).

In addition to the survey, semi-structured face-to-face interviews were undertaken with 15 purposively selected key informants from the hospitality industry in Zimbabwe to ascertain their perception of the concept of sustainability and to identify common sustainable practices by the hospitality sector. An interview guide was used to direct the interview process and permission was sought from interviewees to record the proceedings. Respondents were asked to give their interpretation of sustainability and outline common sustainable practices in the hospitality industry in Zimbabwe. Interviews were analysed using thematic content analysis. Recordings were transcribed verbatim, and open coding was done through an iterative process to give meaning to segments of data. Related codes were further combined into broader themes that explain the popular sentiments from interviewees. The following section presents key findings.

2.4 Sustainability Awareness in the Zimbabwean Hospitality Sector

2.4.1 Survey Results

Table 2.1 captures the major findings of sustainability awareness in the Zimbabwe hospitality sector. It is revealed that overall hospitality businesses in Zimbabwe are conscious of the basic sustainability fact that resources are finite (mean score 3.93). Especially the respondents from the hotel sector realised the limits of resources. The hospitality industry utilises some resources, such as water and energy, very intensively. Therefore, a good awareness of the finite nature of resources is an important ground to understand sustainability thinking, which encourages adopting practices to conserve resources.

In addition, the respondents generally agreed that they are aware of sustainability policies put in place by the regulatory authorities (mean score 3.66). This is an indication that legislation and a comprehensive policy framework are important considerations in building strong awareness and consciousness of sustainability issues in the hospitality industry. This highlights the role of regulatory bodies and industry associations in building a sustainable, conscious hospitality industry. Further, the respondents supported the existence of Standard Operating Procedures (SOPs) in their enterprises to guide employees in the implementation of sustainable practices

Table 2.1 Sustainability awareness in the hospitality sector (N = 125)

Factor	Total sample		Hotels		Lodges		Guest houses	
	N	Mean (SD)	N	Mean (SD)	N	Mean (SD)	N	Mean (SD)
Sustainability in the hospitality industry recognises that resources are finite	122	3.93 (0.929)	21	4.33 (0.913)	47	3.83 (0.985)	54	3.85 (0.856)
My organisation is aware of sustainability policies put in place by the regulatory authorities	123	3.66 (0.895)	21	3.76 (0.944)	47	3.64 (1.009)	55	3.64 (0.778)
My organisation has clearly laid out Standard Operating Procedures (SOPs) to guide employees on sustainable hospitality practices	124	3.60 (1.058)	21	4.05 (1.071)	49	3.53 (1.174)	54	3.50 (0.906)
Sustainable hospitality practices are a criteria used by guests to select facilities to patronise	123	3.37 (1.035)	21	3.71 (1.056)	48	3.35 (1.120)	54	3.26 (0.935)
Employees in my organisation are knowledgeable about sustainable hospitality practices	125	3.26 (1.163)	21	3.52 (1.365)	49	3.12 (1.201)	55	3.27 (1.044)
Guests are willing to pay a premium for sustainable hospitality properties	125	3.21 (1.057)	21	3.52 (1.078)	49	3.10 (1.046)	55	3.18 (1.056)

Note: On a Likert scale 1 = strongly disagree, 2 = Disagree, 3 = Neutral, 4 = Agree, 5 = Strongly agree

(mean score of 3.60). The presence of SOPs in organisations in the hospitality sector in Zimbabwe shows that there is an appreciation that in order to be sustainable, there is an imperative for enacting a set of procedures to guide employee behaviour.

It is indicated that respondents in this study did not believe guests patronising their properties are aware of sustainable practices. This is based on the assertion by respondents that sustainability is not a criterion used by guests to select a property to patronise (a mean of 3.37) and that guests are not willing to pay a premium for sustainable hospitality properties (a mean of 3.21). Moreover, respondents did not perceive their employees are conscious of sustainability issues (a mean of 3.26). This is a notable finding considering that respondents alluded to having SOPs in their organisations to guide employees in sustainable practices. It points to a need to move beyond SOPs to make employees aware of sustainability issues.

2.4.2   Interview Results

The interviews were done with 15 key informants from tourism and hospitality associations and different hospitality organisations (Table 2.2). The purpose of the interviews was to ascertain how sustainability is perceived and understood in the hospitality industry in Zimbabwe and highlight the common sustainable practices. In this regard, the interviewees were asked about their understanding of sustainability as a concept. Two themes emerged from the content analysis of interview transcripts; that sustainability involves preserving the environment for future generations and is centrally about resource conservation.

Table 2.2 Profile of interviewees

Interviewees' code	Organisation	Gender
Interviewee 1	Zimbabwe Tourism Authority (ZTA) official	Male
Interviewee 2	Zimbabwe Tourism Authority (ZTA) official	Female
Interviewee 3	Zimbabwe Tourism Authority (ZTA) official	Female
Interviewee 4	Tourism officer, Ministry of Environment, Tourism and Hospitality	Female
Interviewee 5	Tourism officer, Ministry of Environment, Tourism and Hospitality	Female
Interviewee 6	Tourism Business Council of Zimbabwe (TBCZ) executive	Female
Interviewee 7	Hospitality Association of Zimbabwe (HAZ) executive	Male
Interviewee 8	Hospitality Association of Zimbabwe (HAZ) regional representative	Female
Interviewee 9	Polytechnic lecturer (Harare Polytechnic)	Male
Interviewee10	University lecturer (Great Zimbabwe University)	Male
Interviewee11	University lecturer (Midlands State University)	Female
Interviewee 12	Hotel manager, Harare	Male
Interviewee 13	Lodge manager, Victoria Falls	Male
Interviewee 14	Lodge owner, Victoria Falls	Female
Interviewee 15	Hotel manager, Eastern Highlands	Female

All the interviewees acknowledged that sustainability is premised on the idea of preserving the environment for future generations. This is in line with the basic thinking posited by the World Commission on Environment and Development in the Brundtland report. In this respect, one interviewee indicates that *sustainability refers to ensuring that we maintain the environment for future generations. It is a concept related to living on borrowed time* (Interviewee 1). Similarly, another interviewee stated that *it is about making sure that we do not deplete the environment and its resources because the tourism industry depends on the environment, so the concept of sustainability seeks to ensure that future generations enjoy better than we are today. To me we should leave the world a better place than the state we found it* (Interviewee 9).

The other theme that a significant number of interviewees mentioned was the basic idea that sustainability is all about the conservation of resources. Interviewees noted that the hospitality industry is a major consumer of resources, such as water and energy through various hospitality operations. Interviewees had an option that overuse would result in resource depletion. According to the stakeholder interviewees, the conservation of resources will ensure that the hospitality industry has access to the necessary resources that guarantees its long term survival. One interviewee had this to say regarding resource conservation: *It is about making sure that we do not deplete the environment and its resources because the tourism industry is rooted and dependent on the environment* (Interviewee 10).

2.4.3 Sustainable Hospitality Practices

Overall, based on the thematic content analysis of the sustainable practices by the hospitality industry in Zimbabwe, five broad themes can be discerned. These are the empowerment of local communities, environmentally friendly practices, energy conservation strategies, proper waste management, and water conservation. The need for empowering local communities was stressed by ten of the fifteen interviewees. Broadly, this theme encompasses aspects of Corporate Social Responsibility (CSR) in the Zimbabwe hospitality industry. In practice, it involves buying from local suppliers, introducing indigenous foods on menus, promoting local arts and crafts, and the critical issue of the employment of local people. Such practices benefit the local community as a whole and are based on the thinking that hospitality establishments should make positive contributions to the communities they operate within. Empowering local communities was expressed in the following sentiments by respondents:

There has been focus on foreign foods but I am quite happy that local foods are now available even in the most classy and expensive hotels (Interviewee 10).

If you go to organisation XXX most people employed are locals [.....] and some organisations have been ploughing back their profits to the community (Interviewee 9).

It is evidenced from the above quotes that empowering local communities through including local foods, employing locals and CSR initiatives has a direct impact on changing the livelihoods of local communities and increasing the visibility of the organisation in the community.

The second theme cited by most respondents was the adoption of environmentally friendly practices. This incorporates such activities as green practices, the use of environmentally friendly chemicals, environmentally friendly construction materials as well as joining certification schemes. The hospitality industry is viewed as pursuing such initiatives as a means of protecting the environment it depends on for survival. It can also be observed from the responses that as the carbon footprint by the hospitality industry is significant, practices such as use of environmentally friendly chemicals are regarded as significant for carbon offsetting.

Energy conservation is the core of the sustainability of hospitality businesses as the industry consumes much energy in the kitchens, heating and ventilating guest rooms and for lighting the hospitality property. This makes energy consumption a major cost item that has a considerable impact on the bottom line of hospitality organisations. Interviewees highlighted that in a bid to be sustainable in their operations and to reduce energy costs, certain hospitality organisations use solar energy, energy management systems controlled by room key cards, switch off lights in unoccupied rooms, use of LED lights and procure energy-efficient equipment. Most interviewees mentioned that the favourable climatic conditions in Zimbabwe make it possible to invest in solar energy, thereby reducing electricity bills. Such sentiments were demonstrated by the following observation:

> *Zimbabwe is blessed with sunny weather; we take advantage of that and use solar energy. Because of the electricity situation in the country, generators become costly to use and also emissions to the environment increase* (Interviewee 8).

Interviewees also identified proper waste disposal as a common sustainable practice in the hospitality industry. This is not surprising considering that the hospitality industry is a major producer of solid and liquid waste. Issues highlighted by interviewees include waste separation and recycling. Waste separation was said to be done by using well-labelled rubbish receptacles bins to collect different types of solid waste. Labelling of bins enables the collection of bottles, plastics and paper waste to facilitate recycling. The recycling of food waste and paper was also observed to be a common practice in the hospitality industry. One interviewee expressed the following view:

> *There is waste management through segregation of waste, proper disposal of waste, recycling and banning use of plastic bottles* (Interviewee 13).

Concerning waste management and the general practice of sustainability, another respondent opined: *Some of the common practices by the hospitality industry include recycling, greening of tourism facilities and formation of sustainable management teams* (Interviewee 15).

Lastly, water conservation is a further sustainable practice done by establishments in the hospitality sector of Zimbabwe. The local hospitality industry consumes a lot of water in guest rooms, in the kitchens, watering gardens and in swimming pools. The study revealed that practices such as installation of low flow showerheads, linen reuse policy, bathroom tags emphasising the need to conserve water and recycling of water are common water conservation practices in the Zimbabwe hospitality industry. The interviewees acknowledged that the consumption of water in the hospitality industry is high and therefore, there is a need to adopt strategies that ensure that water is used sparingly. This view is exemplified by the following response:

> The hospitality industry has to identify those areas that use a lot of electricity and a lot of water so that they should look at ways of reducing such consumption (Interviewee 5).

2.5 Discussion of Findings

This study sought to determine sustainability consciousness by the hospitality industry in Zimbabwe. The aim was to probe insights into sustainability awareness, understand how sustainability is perceived and identify the common sustainable practices in the hospitality industry in Zimbabwe. The findings revealed that respondents were in agreement with the assertion that sustainability in the hospitality industry recognise that resources are finite. Indeed, there is mounting evidence in African scholarship that the hospitality sector utilises an extensive amount of natural resources such as water and energy to drive the hospitality business (Khonje et al., 2019). Accordingly, it is not surprising that respondents are aware of the finite nature of resources used by hospitality establishments and of the need to conserve such resources to ensure industry survival.

The research revealed high awareness of sustainability policies put in place by regulatory authorities in Zimbabwe. Non-compliance with sustainability regulations has adverse impacts such as law suits and product boycotts by customers. Hospitality establishments are aware of sustainability policies such as proper waste disposal, use of LED lights and banning of electric geysers, among others so that they are in compliance. Ghaderi et al. (2019) agree that failure to adhere to regulations has the potential to damage the financial and non-financial performance of an organisation as a result of losing environmental conscious customers. In concurrence, Mbasera et al. (2016) also highlighted that governments demand that hospitality establishments adopt sustainable practices hence raising their awareness of current policies they should adhere to. It is, however, clear that awareness of policies put in place by the regulatory authority ensures legitimation and that the business is legally compliant.

Further, there is a consensus among respondents that Standard Operating Procedures (SOPs) exist in their establishments as they are integral in guiding employees on sustainable hospitality practices. Employees' behaviour is an important consideration in achieving a sustainable hospitality organisation (Lombarts, 2018). SOPs help to guide employees and influence positive behavior in the practice of sustainability as it becomes easier for employees to know various ways they can implement sustainable practices. In South Africa, Rogerson and Sims (2012) showed that awareness programmes are important in order to educate staff on their role in practising sustainability. Indeed, codes of conduct are seen as important guidelines which help organisations and their members to commit to sustainable practices (Weaver et al., 2013). Setting guidelines for employees to follow is an effective way of raising sustainability awareness among employees in Zimbabwe.

Generally, respondents were not in agreement with the assertions that sustainability is a criteria used by guests to select a hospitality facility to stay nor that guests would be willing to pay a premium for sustainable hospitality facilities. Also, they disagreed that employees in their organisations are aware of sustainability issues. Njerekai (2019) noted that the low level of awareness by employees on sustainability issues is credited to the absence of sustainability policies in certain hospitality establishments, which leads to inconsistent application of sustainable practices. With regards to guest awareness, the findings of this research contrast with De Freitas (2018) work on intended behaviour towards selecting green hotels by South African consumers, which revealed that customers tend to search for green properties and are sometimes willing to pay more for such facilities.

The findings of this study further demonstrated that sustainability awareness is much high in hotels as compared to lodges and guest houses. This can be attributed to the difference in scale and size of their operations. Hotels operate at a large scale as compared to lodges and guests houses; therefore, their impact on the environment, economy and society are visible, prompting them to adopt sustainable practices. In addition, sustainability initiatives require a significant financial commitment, which usually is unavailable in small accommodation service establishments such as lodges and guest houses. Agyeiwaah (2019) acknowledges the leading role of hotels in embracing sustainability and noted that small operators lag behind in that regard. From a research base in the global North, Kornilaki et al. (2019) signal that engagement in sustainability initiatives by SMEs in the tourism industry is low variously because of lack of awareness of required actions, unavailability of skills to implement sustainable practices, lack of interest from customers and the general public, and because of resource constraints.

Results from interviews highlighted that sustainability in the hospitality industry in Zimbabwe is seen as a concept that furthers the agenda of protecting the environment for future generations and focuses on ensuring resource conservation. This viewpoint aligns with the definition of sustainable development as posited by the World Commission on Environment and Development (WCED) in 1987. The findings also reveal that interviewees are aware that sustainable hospitality involves "managing resources considering the economic, social and environmental costs and benefits in order to meet the needs of the present generations while protecting and enhancing opportunities for future generations" as proposed by (Legrand & Nielsen, 2017). Likewise, Worku and Mohammed (2019) from Ethiopian research point out that sustainability is perceived as conservation and proper use of natural resources in a way that enables future generations to use the same resources. It is clear from the findings of this study that the interviewees were aware of the principles guiding sustainability in the hospitality sector.

Specific sustainable initiatives highlighted by interviewees include CSR, charity donations, green certification, environmental considerations in construction, use of LED lights, waste separation, recycling, linen reuse policy and low flow shower-heads. These findings corroborate the results of research undertaken at Victoria Falls, Zimbabwe, by Dube and Nhamo (2020), which stresses that the hospitality industry is implementing various green initiatives to reduce its carbon footprint. Similarly, Njerekai (2019) observed that the linen and towel reuse policy, waste separation, use of guest key cards to control in-room energy use and green certifica-tion are some of the sustainable practices adopted by the Zimbabwe hospitality industry. In a review of research undertaken in Iran Ghaderi et al. (2019) concurred with this study that the hospitality industry engages in community involvement through supporting charities and the underprivileged members of the communities, employing locals, and contributing to the preservation of local cultures. In scrutinis-ing sustainable certification schemes in tourism and hospitality Mzembe et al. (2020) further agreed that hospitality organisations exploit environmental and social resources for economic gains. Thus the need exists to ensure that they shoulder cor-responding costs to reduce resource use.

2.6 Conclusion

This study represents a modest contribution to sustainability debates surrounding the hospitality sector in the global South and specifically to the context of sub-Saharan Africa. It must be concluded from this research that the hospitality sector in Zimbabwe is aware of the principles and practices that guide the concept of sustain-ability. The key informants demonstrated that they are aware of the salient features of a sustainable hospitality establishment. Hospitality operators in Zimbabwe are aware that they use finite resources in their operations and aware of sustainable poli-cies put in place by national authorities to regulate the industry. Policies and regula-tions are effective in enforcing sustainability as the industry takes note of their existence. Moreover, it can be concluded the availability of Standard Operating Procedures (SOPs) to guide employees in the practice of sustainability is a signal that the hospitality sector takes sustainability issues seriously such that it is believed there is a need to institutionalise them and guide employees on how to act. Nevertheless, the general low levels of awareness among employees on sustainabil-ity issues (regardless of the availability of SOPs) is an indication that there is a need for initiatives that go beyond guidelines to ensure that employees are conscious of sustainability issues. Training, education programmes and incentives may be neces-sary to motivate employees to take part in sustainability activities in Zimbabwe hospitality establishments. The study revealed that customers do not use sustain-ability issues as criteria to select and book a property and are unwilling to pay a premium for sustainable hospitality properties signifies that sustainability consciousness among them is low. Various sustainable practices were said to be implemented by the hospitality industry. These include CSR, local employment, buying from locals, proper waste disposal, joining certification schemes, use of environmentally friendly chemicals, energy and water conservation strategies.

Given that awareness of sustainability practices among employees and customers is observed to be low, further investigations are required concerning customer preferences to understand better the sustainable practices they are interested in. In addition, studies are needed on employee attitudes towards sustainability practices to gauge how their behavioural intentions can be influenced positively. Overall, awareness of sustainable issues makes operators and employees receptive to sustainable practices, industry initiatives as well as global initiatives such as SDGs to achieve a sustainable and more socially inclusive hospitality industry. Sustainable practices by the hospitality sector could result in cost savings, appeals to environmental conscious customers, improve the image of the organisation, inclusivity, conserve resources and achieve legitimation, among other benefits. Therefore, sustainability conscious hospitality operators are in the central position governing the hospitality industry towards a sustainable development path in Zimbabwe.

References

Agyeiwaah, E. (2019). Exploring the relevance of sustainability to micro tourism and hospitality accommodation enterprises (MTHAEs): Evidence from home-stay owners. *Journal of Cleaner Production, 226*, 159–171. https://doi.org/10.1016/j.jclepro.2019.04.089

Ayuso, S. (2007). Comparing voluntary policy instruments for sustainable tourism: The experience of the Spanish hotel sector. *Journal of Sustainable Tourism, 15*(2), 144–159.

Beusch, P., & Björnefors, E. (2014). *Hospitality and sustainability: A case-study and comparison of the sustainability work by hotels in the Gothenburg area background.* University of Gothenburg.

Bennet, A., & Bennet, D. (2008). Moving from knowledge to wisdom, from ordinary consciousness to extraordinary consciousness. *Vine, 38*(1), 7–15. https://doi.org/10.1108/03055720810870842

Bianchi, R. (2020). COVID-19 and the potential for a radical transformation of tourism? *ATLAS Tourism and Leisure Review, 2020-2*, 80–86.

Chiutsi, S., & Saarinen, J. (2019). The limits of inclusivity and sustainability in transfrontier peace parks: Case of Sengwe community in Great Limpopo Transfrontier Conservation Area, Zimbabwe. *Critical African Studies, 11*(3), 348–360. https://doi.org/10.1080/2168139 2.2019.1670703

Claver-Cortés, E., Molina-Azorín, J. F., Pereira-Moliner, J., & López-Gamero, D. (2007). Environmental strategies and their impact on hotel performance. *Journal of Sustainable Tourism, 15*(6), 663–679. https://doi.org/10.2167/jost640.0

Coles, T., Fenclova, E., & Dinan, C. (2013). Tourism and corporate social responsibility: A critical review and research agenda. *Tourism Management Perspectives, 6*, 122–141. https://doi.org/10.1016/j.tmp.2013.02.001

Davis, K. (1960). Can business afford to ignore social responsibilities? *California Management Review, 2*(3), 70–77.

De Carvalho, B.L., de Fatima Salgueiro, M., & Rita, P. (2015). Consumer sustainability consciousness: A five dimensional construct. *Ecological Indicators, 58*, 402–410. https://doi.org/10.1016/j.ecolind.2015.05.053

De Freitas, D. (2018). *Exploring and predicting south African consumers' intended behaviour towards selecting green hotels: Extending the theory of planned behaviour.* University of South Africa.

Dodds, R., & Kuehnel, J. (2008). CSR among Canadian mass tour operators: Good awareness but little action. *International Journal of Contemporary Hospitality Management, 22*(2), 221–244.

Dube, K., & Nhamo, G. (2020). Greenhouse gas emissions and sustainability in Victoria Falls : Focus on hotels, tour operators and related attractions greenhouse gas emissions and sustainability in Victoria Falls. *African Geographical Review.* https://doi.org/10.1080/1937681 2.2020.1777437

Elkington, J. (1998). *Cannibals with forks: The triple bottom line of 21st century business.* New Society.

Franzoni, S., & Avellino, M. (2019). Sustainability reporting in international hotel chains. *Symphonya Emerging Issues in Management, 1*, 96–107.

Friedman, M. (1970, September 13). The social responsibility of business is to increase its profits. *New York Times Magazine*, 122–124. Retrieved from http://umich.edu/~thecore/doc/ Friedman.pdf

Ghaderi, Z., Mirzapour, M., Henderson, J. C., & Richardson, S. (2019). Corporate social responsibility and hotel performance: A view from Tehran, Iran. *Tourism Management Perspectives, 29*, 41–47. https://doi.org/10.1016/j.tmp.2018.10.007

Hambira, W., Saarinen, J., Atlhopheng, J., & Manwa, H. (2013). Climate change adaptation practices in nature-based tourism in Maun in the Okavango Delta area, Botswana: How prepared are the tourism businesses? *Tourism Review International, 17*(2), 19–29.

Hobson, K., & Essex, S. (2001). Sustainable tourism: A view from accommodation businesses. *The Service Industries Journal, 21*(4), 133–147.

Huang, H., Lin, T., Lai, M., & Lin, T. (2014). Environmental consciousness and green customer behavior: An examination of motivation crowding effect. *International Journal of Hospitality Management, 40*, 139–149. https://doi.org/10.1016/j.ijhm.2014.04.006

Khonje, L. Z., Simatele, M. D., & Musavengane, R. (2019). Environmental sustainability innovations in the accommodation sub-sector: Views from Lilongwe, Malawi. *Development Southern Africa*, 1–16. https://doi.org/10.1080/0376835x.2019.1660861

Kimaro, E., & Saarinen, J. (2020). Tourism and poverty alleviation in the Global South: Emerging corporate social responsibility in the Namibian nature-based tourism industry. In M. Stone, M. Lenao, & N. Moswete (Eds.), *Natural resources, tourism and community livelihoods in southern Africa* (pp. 123–142). Routledge.

Kornilaki, M., Thomas, R., & Font, X. (2019). The sustainability behaviour of small firms in tourism: The role of self-efficacy and contextual constraints. *Journal of Sustainable Tourism, 27*(1), 97–117. https://doi.org/10.1080/09669582.2018.1561706

Lee, J.-S., Hsu, L.-T., Han, H., & Kim, Y. (2010). Understanding how consumers view green hotels how a hotel s green image can influence behavioural intentions. *Journal of Sustainable Tourism, 18*(7), 901–914. http://www.tandfonline.com/action/showCitFormats? doi=10.1080/09669581003777747

Legrand, W., & Nielsen, R.S. (2017). Climate conscious identity and climate-adaptive innovations in hospitality. *Advances in Hospitality and Leisure, 13*, 63–78. https://doi.org/10.1108/ S1745-354220170000013006

Lombarts, A. (2018). The hospitality model revisited: Developing a hospitality model for today and tomorrow. *Hospitality & Society, 8*(3), 297–311. https://doi.org/10.1386/hosp.8.3.297_7

Low, A. (2005). What is consciousness and has it evolved? *World Futures, 61*(3), 199–227. https:// doi.org/10.1080/026040290503423

Mbasera, M., Du Plessis, E., Saayman, M., & Kruger, M. (2016). Environmentally-friendly practices in hotels. *Acta Commercii, 16*(1), 1–8. https://doi.org/10.4102/ac.v16i1.362

Middleton, V. T. C., & Hawkins, R. (1998). *Sustainable tourism: A marketing perspective.* Butterworth Heinemann.

Millar, M., Mayer, K. J., & Baloglu, S. (2012). Importance of green hotel attributes to business and leisure travelers. *Journal of Hospitality Marketing & Management, 214*(10), 1936–8623. https://doi.org/10.1080/19368623.2012.624294

Minoli, D. M., Goode, M. M. H., & Smith, M. T. (2015). Are eco labels profitably employed in sustainable tourism? A case study on Audubon Certified Golf Resorts. *Tourism Management Perspectives, 16*, 207–216. https://doi.org/10.1016/j.tmp.2015.07.011

Musavengane, R. (2019). Understanding tourism consciousness through habitus: Perspectives of ' poor ' black South Africans. *Critical African Studies, 11*(3), 322–347. https://doi.org/10.108 0/21681392.2019.1670702

Mzembe, A. N., Lindgreen, A., & Idemudia, U. (2020). A club perspective of sustainability certification schemes in the tourism and hospitality industry. *Journal of Sustainable Tourism, 28*(9), 1332–1350. https://doi.org/10.1080/09669582.2020.1737092

Njerekai, C. (2019). Hotel characteristics and the adoption of demand oriented hotel green practices in Zimbabwe: A regression. *African Journal of Hospitality, Tourism and Leisure, 8*(2), 1–16.

Prud'homme, B,. & Raymond, L. (2016). Implementation of sustainable development practices in the hospitality industry. *International Journal of Contemporary Hospitality Management, 28*(3), 609–639. https://doi.org/10.1108/IJCHM-12-2014-0629

Radwan, H. R. I., Jones, E., & Minoli, D. (2010). Managing solid waste in small hotels. *Journal of Sustainable Tourism, 18*(2), 175–190. https://doi.org/10.1080/09669580903373946

Ricaurte, E. (2011). Developing a sustainability measurement framework for hotels: Toward an industry-wide reporting structure. *Cornell Hospitality Report, 11*(13), 6–30.

Rogerson, C. M., & Rogerson, J. M. (2011). Tourism research within the southern African development community: Production and consumption in academic journals, 2000-2010. *Tourism Review International, 15*(1–2), 213–224.

Rogerson, J. M., & Sims, S. R. (2012). The greening of urban hotels in South Africa: Evidence from Gauteng. *Urban Forum, 23*(3), 391–407. https://doi.org/10.1007/s12132-012-9160-2

Rogerson, J. M., & Visser, G. (Eds.). (2020). *New directions in South African tourism geographies.* Springer.

Russo, M. V., & Fouts, P. A. (1997). A resource-based perspective on corporate environmental performance and profitability. *Academy of Management Journal, 40*(3), 534–559.

Saarinen, J. (2018). Beyond growth thinking: The need to revisit sustainable development in tourism. *Tourism Geographies, 20*(2), 337–340. https://doi.org/10.1080/14616688.2018.1434817

Saarinen, J. (2019). Communities and sustainable tourism development: Community impacts and local benefit creation tourism. In S. F. McCool & K. Bosak (Eds.), *A research agenda for sustainable tourism* (pp. 206–222). Edward Elgar Publishing.

Saarinen, J. (2020). Tourism and sustainable development goals: Research on sustainable tourism geographies. In J. Saarinen (Ed.), *Tourism and sustainable development goals: Research on sustainable tourism geographies* (pp. 1–10). Abington.

Saarinen, J., Hambira, W., Atlhopheng, J., & Manwa, H. (2012). Perceived impacts and adaptation strategies of the tourism industry to climate change in Kgalagadi South District, Botswana. *Development Southern Africa, 29*(2), 273–285.

Scheyvens, R. (2018). Linking tourism to the sustainable development goals: A geographical perspective. *Tourism Geographies, 20*(2), 341–342. https://doi.org/10.1080/1461668 8.2018.1434818

UNWTO (World Tourism Organization). (2005). *Making tourism more sustainable - A guide for policy makers.* UNWTO.

WCED. (1987). *Report of the world commission on environment and development: Our common future.* United Nations, New York.

Weaver, D., Davidson, M. C. G., Lawton, L., Patiar, A., Reid, S., & Johnston, N. (2013). Awarding sustainable Asia-Pacific hotel practices: Rewarding innovative practices or open rhetoric? *Tourism Recreation Research, 38*(1), 15–28. https://doi.org/10.1080/02508281.2013.11081726

Worku, Z., & Mohammed, T. (2019). Eco-lodges and tourist infrastructure development in and around Abijata Shalla Lakes National Park: From the perspective of evaluating their sustainability. *Journal of Tourism, Hospitality and Sports, 45*. https://doi.org/10.7176/JTHS/45-02

World Business Council for Sustainable Development. (2002). Making a difference toward the Johannesburg Summit 2002 and beyond. *Corporate Environmental Strategy, 9*(3), 26–235. https://doi.org/10.016/S0667938(02)000714

Ngoni Courage Shereni is a faculty member at Lupane State University in Zimbabwe. Currently, he is pursuing a PhD in Tourism and Hospitality at the University of Johannesburg in South Africa. His research interests are in sustainable tourism, Sustainable Development Goals (SDGs) in the Tourism and Hospitality industry, tourism and hospitality education, disruptive technology in the tourism industry, tourism exhibitions as well as Community-based Natural Resources Management (CBNRM) practices, among others.

Jarkko Saarinen is a Professor of Human Geography (Tourism Studies) at the University of Oulu, Finland, and Distinguished Visiting Professor (Sustainability Management) at the University of Johannesburg, South Africa, and Extraordinary Professor at the Tourism Management Division, Department of Marketing Management, University of Pretoria. His research interests include sustainable development, sustainable tourism, tourism-community relations and nature conservation studies.

Christian M. Rogerson is a Research Professor at the School of Tourism & Hospitality, University of Johannesburg, South Africa. His research interests focus on the nexus of tourism and development and specifically issues of local economic development and tourism small business development. Among his recent publications is the co-edited volume on Urban Tourism in the Global South: South African Perspectives (Cham, Switzerland; Springer Nature, 2021).

Chapter 3
In Pursuit of Sustainable Tourism in Botswana: Perceptions of Maun Tourism Accommodation Operators on Tourism Certification and Eco-Labelling

Godiraone Trompies Motsaathebe and Wame L. Hambira

3.1 Introduction

Since the advent of sustainable development dialogue spanning momentum as far back as 1992 with the Earth Summit that led to the Rio Declaration on Environment and Development, many sectors, including tourism, have aligned themselves to this discourse. Sustainable tourism is, therefore, a product of the sustainable development philosophy in recognition of the impacts of tourism on the environment and socio-cultural aspects of human wellbeing. Undeniably, tourism is responsible for 'annual human migration', leading to negative impacts on the natural environment, economies, and livelihoods (Blackman et al., 2014; Budeanu, 2005). Thus, sustainable tourism is defined as tourism with practical intents to reduce environmental and socio-cultural impacts (Buckley, 2002). It is therefore anchored in nature-based tourism in which the natural environment is the principal component of the product or activity and through ecotourism whose strength lies in environmental education and conservation (Buckley, 2002).

Due to the complex relationship between tourism development and environmental protection, Ruhanen et al. (2015) point out that research focus on sustainable forms of tourism has increased considerably. Consequently, several approaches have been developed over time to facilitate sustainable tourism, and these include Environmental Management Systems (mechanisms adopted by companies to manage their environmental matters in a systematic and holistic manner); Local Agenda 21 (where businesses collaborate with communities to develop strategies aimed at shaping local programs and policies towards sustainable development using the principles of the Agenda); Cleaner Productions (strategies towards processes that

G. T. Motsaathebe · W. L. Hambira (✉)
University of Botswana Okavango Research Institute, Maun, Botswana
e-mail: whambira@ub.ac.bw

J. Saarinen et al. (eds.), *Southern African Perspectives on Sustainable Tourism Management*, Geographies of Tourism and Global Change,
https://doi.org/10.1007/978-3-030-99435-8_3

reduce the use of natural resources and prevention of pollution); and eco-labelling (tools used to inform consumers that certain products or services have met agreed environmental performance standards) (Lee, 2001). Therefore, the focus of this chapter is on eco-labels as applied to the tourism accommodation sector in Botswana.

Tourism eco-labels have been developed as one of the tools to foster sustainable tourism (Jensen et al., 2004). They are used as mechanisms to reduce environmental impacts and gain competitive advantage (Font & Harris, 2004). Eco-labelling is defined as an environmental performance certification system, and eco-labels identify products and services that are proven to be environmentally friendly (Global Ecolabelling Network, 2016). A best practice is measured by energy and water efficiency as well as minimization of waste and wastewater production (UNEP, 1998 cited in Warnken et al., 2005). According to Bratt et al. (2011), the purpose of eco-labels is based on perspectives of the producer, the consumer, and the policymaker. To the producer, the eco-label is regarded as an instrument that shows the environmental and social performance of products and services; to the consumer, it is a source of information on the quality of products and services; and to the policymaker, it is seen as a complementary instrument for the creation of incentives for innovation of products with less impact on the environment (Bratt et al., 2011). The consumer is, however, key in the whole process since many tourism businesses and destinations believe firmly that consumers' choice is driven by motives with a background consideration for the environment (Font & Buckley, 2001). This belief, amongst other reasons, has led to the establishment of a vast number of local, regional, and national tourism eco-labelling schemes in order to promote environmentally friendly tourism businesses.

Tourism research in Botswana has also shown that the tourism industry has negative impacts on the environment, and tools such as proper certification and eco-labels could be a solution to achieve sustainable tourism practices (Mbaiwa, 2002; Mbaiwa et al., 2011; Moswete et al., 2019). Consequently, in 2010, the Botswana Tourism Organisation (BTO) introduced the Botswana Ecotourism Certification System (BECS) as one of the efforts towards the development of sustainable tourism in the country. Since then, a few accommodation facilities such as the Grand Palm Hotel Casino and Convention Centre (now Walmont Hotel) and Chobe Game Lodge have participated in the programme (Grand Palm Resort, 2016; Chobe Game Lodge, 2016). This chapter, therefore, aims to determine the Maun tourism operators' perceptions about tourism eco-labelling as a tool to achieve tourism sustainability and help to minimize the environmental impacts of their operations. The research, therefore, seeks to answer the questions: How do tourism businesses perceive the tourism eco-labelling scheme, and what are the reasons behind the seemingly slow uptake of the program?

3.2   Tourism Certification and Eco-Labels: A Review

A tourism eco-label is a certification obtained by a tourism company in recognition of attempts made to reduce the impact of its operations on the environment; hence their content refers mainly to the environment (Buckley, 2002). In tourism, the concept of eco-labelling emanates from the global concern about the negative environmental impacts of the tourism industry. Its roots are found in various disciplines such as green economy, ecotourism, natural capital accounting, marketing, and environmental economics (Mbaiwa et al., 2011; Petan et al., 2007; Pieterse, 2004). Eco-labels have attracted the attention of international tourism and environmental organizations with the greatest proliferation of tourism eco-labels being in Europe mainly for marketing purposes (Buckley, 2002). They were introduced in the 1980s and have since spread to developing countries (Blackman et al., 2014) such as Botswana. Even then, eco-labels have been developed in ununiformed patterns, and in most parts of the developing countries, there are still no eco-labelling schemes in operation (Pieterse, 2004). The certification schemes are voluntary and often promoted under the belief that the conventional command and control instruments have failed or are ineffective (see Berghoef & Dodds, 2013).

According to Githinji (2006), potential benefits of tourism eco-labels to tourism enterprises include but are not limited to: reduction of operational costs through increased process efficiency, improvement of environmental performance and promotion/marketing of companies participating in the programme. Tourism eco-labelling is, thus, not just a certification program reward but also a strong marketing tool that can help the tourism business to practice good environmental practices whilst still promoting their business at the same time. Mbaiwa et al. (2011) further opine that certification has the potential to reduce tourism's negative environmental and social impacts, not only through the setting of performance standards but also by ensuring that the tourism industry is held accountable to stakeholders, including destination communities. Githinji (2006) equally notes that tourism eco-labels aim to inform consumers about the environmental impacts of the products consumed and therefore could be having some form of influence on the way tourists choose to visit a particular tourism destination or stay at a particular tourism accommodation facility. Eco-labels are therefore aimed at: communicating information about the tourism facility, which the customer can use to make informed decisions; notifying customers of the environmental impacts of the tourism facility to encourage a switch towards sustainable resorts; and conveying positive messages about the quality of environmental services offered by the tourism entity (see Minoli et al., 2015). Subsequently, 'consumers are able to reward those who are participating through purchasing behaviour and likewise to 'punish' those who do not' (Berghoef & Dodds, 2013, p.264; see Blackman et al., 2014).

However, tourism eco-labelling certification schemes around the world are still faced with several challenges that hinder their successful implementation and uptake by tourism businesses. The schemes have numerous limitations, such as influencing consumer choice in selecting a tourism destination (Kozak & Nield,

2004). Another great challenge that still persists, as highlighted by Font (2002), is that there are too many tourism eco-labels with different meanings which rely on governments for funding, hence their limited target groups and inability to grow. Furthermore, the schemes have been observed with suspicion with regards to the motives behind their uptake by companies as well as limited participation by companies which is attributed to, among other things, financial constraints; enforcement challenges due to their voluntary nature; negative public perception because of poor execution and planning (Berghoef & Dodds, 2013). Moreover, eco-labelling schemes have been criticized for having a strong environmental focus whereas sustainability is not only about ecological aspects but also about social and economic impacts of the tourism industry (Tepelus & Cordoba, 2005). This, however, is said to be attributed to the fact that the schemes are mainly aimed at reducing costs related to utilities such as energy and water, whereas it is difficult to attach a price to the integration of socio-cultural aspects into their operations (Tepelus & Cordoba, 2005).

3.3   Tourism Certification and Eco-Labelling in Botswana

The government of Botswana has long subscribed to the sustainable development discourse as evidenced by the ratification of multilateral environmental agreements that promote sustainable development, such as the Convention on Biological Diversity and the United Nations Framework Convention on Climate Change. Furthermore, the quest for sustainable tourism can be traced back to the 1990 Botswana tourism policy and the national ecotourism strategy of 2003. Thus, the introduction of tourism eco-certification was an extension of the culture of sustainable development path that the Botswana government has adopted as a member of the international community.

In the year 2010, the Botswana Tourism Organisation (BTO, 2010), a parastatal corporate body established by the government to market the Botswana tourism products, grade and classify tourism accommodation facilities, introduced the Botswana Ecotourism Certification System (BECS). This was in addition to the star-rating system for quality assurance in the accommodation facilities. BTO (2010) defines BECS as a voluntary industry program that covers more than 240 performance standards encompassing environmental management, cultural resources protection and community development, socio-economic responsibility and the fundamental ecotourism criteria. The BECS comprises of accommodation and Eco-tour standards (see Mbaiwa et al., 2011). Standards are aimed at achieving a certain level of environmental performance without which companies will not be able to comply and ultimately resort to their discretion hence defeating the purpose (Font, 2002). Performance standards, therefore, mean that every company receiving an eco-label would have met the pre-specified thresholds (Font, 2002).

Mbaiwa et al. (2011) observe that the BECS was adopted to facilitate conservation and sustainability within its rapidly growing tourism industry, with the most target areas being the Okavango Delta and Chobe regions. The BECS was also developed in line with Botswana's High Value -Low Volume tourism strategy to conserve its natural environment. BTO (2010) also created a specific manual for ecotourism businesses to assist them in implementing eco-friendly business operations. Some of the examples of environmentally friendly operations that certified operators should adhere to include but are not limited to; constructing their facilities with timber from sustainable certified forests, using alternative power sources like solar panels, chlorine-free swimming pools, planting of indigenous vegetation and the use of low energy consumption power appliances. All these practices help businesses minimize the negative environmental impacts of their tourism facilities.

3.4   Methods

3.4.1   Study Area

This study was carried out in Maun, the tourism hub of Botswana and the gateway to the internationally acclaimed Okavango Delta and the Moremi Game Reserve located in the northwestern part of Botswana (see Fig. 3.1). Maun is the administrative capital of the Ngamiland District, and according to the Central Statistics Office's 2011 population census, it had a population of 55,784. Maun was chosen as the study area since it is the centre for most of the tourism businesses found in the

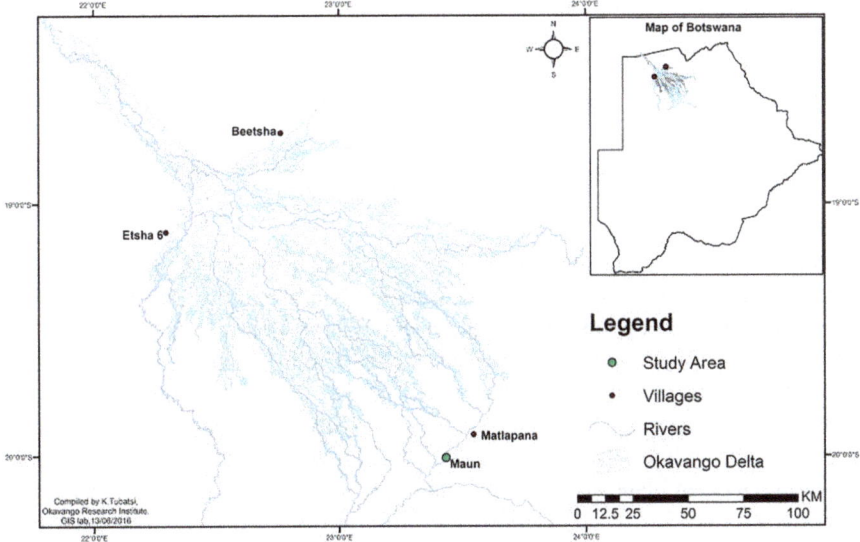

Fig. 3.1 Map showing study area. (Source: Okavango Research Institute, GIS Lab)

Ngamiland region and was also targeted during the development of the BECS (see Mbaiwa et al., 2011).

3.4.2   Data Collection and Analysis

The research adopted a qualitative approach targeting tourism businesses offering accommodation services in Maun operating under the tourism licence A or B, which are the target for the Botswana Tourism Organisation's BECS. The accommodation Standards category A are accommodation facilities on a fixed site comprising hotels, camps, guest houses (including corporate guest houses), bed and breakfast, lodges, backpacker tourist accommodation, and self-catering facilities outside protected areas, as well as cultural villages and timeshare facilities. Category B comprises accommodation on fixed sites but offers game drives and other outdoor activities within wildlife management areas as well as protected areas. These include lodges, photographic camps, hunting camps, campsites and timeshare facilities. (Botswana Investment and Trade Centre, 2016).

Preliminary data was collected using semi-structured interview schedules, which were administered to tourism corporate executives in order to determine their awareness and perceptions about eco-labelling and their views regarding its uptake by the Botswana's tourism industry. According to Bernard (1988), semi-structured interviewing is best suited in instances where the interviewer will not get more than one chance to interview someone. This study targeted corporate executives who are usually very busy to accord time for interviews hence the preference for semi-structured interviews. This method also allows interviewees to express their opinions freely and in detail while the interviewer can make follow up questions where needed (see Partington, 2001). The researcher also observed if there were any signs that support the application of the principles of sustainable tourism by the establishment, such as the environmentally friendly features addressing aspects of energy and waste management. Additional data was collected through both published and unpublished documents such as journals articles and government documents that addressed sustainable tourism in general, as well as tourism certification and eco-labelling matters.

A purposive sampling approach was used to identify tourism businesses offering accommodation services in Maun based on their proximity and accessibility. Polkinghome (2005), points out that the selection of interview participants in qualitative research requires the use of a purposive sampling method to gain in depth understanding of the situation. In addition, limited time and resources were also considered in choosing the sampling method. A list of all licensed tourism businesses operating under categories A and B was obtained from the Department of Tourism (DOT) in Maun and used as a sampling frame. The DOT list was used hand in hand with the list of all facilities that have attained any certain level of eco-certification grade as obtained from BTO website. In the end, 19 companies representing 47 facilities (some companies intervicwed owned more than one establishment) offering accommodation services participated in this study. An interpretative approach emphasizing determining patterns, categories and basic descriptive units (Kitchin & Tate, 2000) was used to analyze the data. This entailed: reading through all responses recorded in all the interview schedules; labelling of responses according to recurring responses; categorizing them according to themes; and interpretation of key information.

3.5   Results and Discussions

3.5.1   Background Information of Interviewees and Business Entities they Represented

The interviewees comprised of representatives of the 19 companies (hotels, camps, guesthouse, lodge, self-catering, bed & breakfast, and campsites) operating under the DOT's A or B licence in order to gain insight of their perceptions towards tourism eco-labelling. Of the tourism companies represented, 57.9% of their facilities were classified under category A and the remaining 42.1% operated under the B licence category. The tourism businesses that were contacted for an interview had both of their registered office and facility in Maun or only the registered office located in Maun while the facility was in the Delta. All interviewees were in executive positions such as supervisors (42.1%), general managers (21.1%), operations managers (5.8%), human resource managers (10.5%) and environmental managers (10.5%). Environmental managers constituted experts on environmental issues and were only present in big tourism companies which were eco-certified. Of all the interviewed companies, only 10.5% had some of their facilities eco-certified, while the remaining 89.5% of them had none of their facilities eco-certified. The majority of the interviewed facilities were camps (36.8%), and two of them had attained a certain level of eco-certification grade (Table 3.1). The two eco-certified camps are located within the Wildlife Management Areas (WMAs) in the Okavango Delta.

Table 3.1 Tourism businesses that participated in the study

	Frequency	Per cent
Hotel	4	21.1
Camp	7	36.8
Guest house	1	5.3
Lodge	3	15.8
Self-catering	1	5.3
Bed & breakfast	1	5.3
Campsite	2	10.5
Total	**19**	**100.0**

Source: Field Survey, 2016

Participation of local people in the tourism industry is one of the core principles behind tourism certification, such as the BECS program. The majority (89%) of the business representatives interviewed were Batswana, indicating that most of these tourism operators employed local staff even in the higher management positions (see Moswete et al., 2019). This, however, should not be construed to be equivalent to ownership as Mbaiwa (2015) has already observed that almost 80% of tourism businesses in the Okavango Delta are foreign owned or have a direct influence of foreign ownership.

3.5.2 Tourism Eco-Labelling: Perceptions of Maun Tourism Operators

Although the concept of tourism certification was embraced by almost all the interviewees, most of them (63.2%) had little to no knowledge about the BECS program. This lack of knowledge about the program may therefore have a bearing on the way in which tourism operators perceive it and their willingness to participate in it. Perceptions varied between the eco-certified and un-certified businesses, but almost all the tourism businesses were of the view that tourism eco-labelling is a good initiative that can help to minimize the negative environmental impacts of tourism and promote sustainable tourism development. One of the respondents who had one of their facilities eco-certified mentioned that 'BECS is a good program but not realistic because it is hindered by lack of infrastructure to accommodate it'. Another one emphasized that BECS is 'a very good platform for the tourism industry, in general, to manage their businesses in the most eco-friendly manner'.

Those that were not certified also believed that being BTO eco-certified would positively influence their operations towards environmental sustainability. This may have been influenced by the fact that eco-friendly practices such as energy-saving practices, proper waste management and recycling form the core part of the BECS accommodation standards, which all certified facilities should adhere to as prescribed in the Botswana Ecotourism Best Practices Manual (Botswana Tourism Board, 2002). The examples of environmentally friendly operations that certified operators should adhere to help the business to minimize the negative environmental impacts of their tourism facilities. Those tourism businesses that never participated in the BECS program pointed out that they think it is a good program despite several challenges that hinder them from participating in it. One of them stated that 'It's a good system despite the fact that it requires funding\money to match'.

3.5.3 Tourism Eco-Certification and Eco-Labelling: Experiences from Maun

The companies which were BTO eco-certified similarly mentioned ways in which they benefited from being eco-certified, mainly a marketing advantage. This was especially emphasized where international tourists formed most of the visitors since, according to the interviewee, being BTO certified meant they receive free marketing in the international scene. In order to promote the BECS, BTO has partnered with other globally recognized organizations such as Fair-Trade Tourism in South Africa (Fair trade tourism, 2016). This is in line with what Pieterse (2004) observed that internationally accredited eco-labels are bound to enable the tourism businesses to compete in the global scene. According to the operators, BTO eco-labels are still new and not well known in the international tourism arena; hence this kind of partnership will help to promote their brands and boost their business. Since tourism is a globally recognized industry, it requires eco-labels that are globally recognizable (Buckley, 2002).

The interviewees declared that being certified came with strict requirements that a facility must adhere to while at the same time encouraging them to attain the next higher level in sustainable practices. They also highlighted that BTO carried out frequent visits to their facilities to ascertain whether they complied with the standards, and this encouraged them to always operate as expected. The study revealed that some of the respondents (52.8%), who were mostly big tourism operators are of the view that tourists do consider tourism eco-labels when selecting their tourism destinations. Some of the interviewees, for example, mentioned that they sometimes receive inquiries such as 'How eco-friendly is your business?' from some tourists when making accommodation bookings at their facilities. Such inquiries indicate that some tourists do recognize the need to contribute towards sustainable tourism and the green economy. Tourism certification thus acts as an assurance to the consumer that the business is being run in an environmentally sustainable manner. Hence, for a lodging facility to not be certified has a high potential to disadvantage the tourism entity. Indeed, consumers (tourists) play a key role in the success of eco-labelling and therefore, their opinion should be taken seriously. For example, according to Prieto-Sandoval et al. (2016), the eco-labelling innovation cycle commences with consumers expressing their environmental expectations leading to companies improving their existing products to satisfy these expectations, which in turn increases the consumers' expectations, and the cycle starts again. However, one of the representatives of establishments that were certified were of the view that it was still not clear if tourists who visited their facilities were really motivated by the fact that they had a BTO eco-certification logo while one of the interviewees said, 'It is too early to judge if our company really benefit from being eco-certified.'

Related to the issue of awareness, the demand by tourists as mentioned above may indicate the growth in awareness of eco-labelling by international tourists since other earlier studies such as Leonard (2011) found out that Austrian tourists are not yet knowledgeable about tourism sustainability and may not be willing to pay more for certified tourism products. In the context of Botswana, a clear validation of this aspect could have been possible only if tourists had been interviewed to get their direct view concerning their tourist destination decisions and eco-labels. This was, however, outside the scope of this study.

3.5.4 Perceptions on Reasons Behind the Slow Uptake of Tourism Eco-Labelling Schemes by Tourism Businesses

On what could be the reason for the slow uptake of the BECS program, several factors were identified, with lack of awareness being the most prominent factor (57.9%). Other factors highlighted were lack of skills (15.8%), financial constraints (26.3%), poor government supporting infrastructure (5.2%) and incompatible business culture (5.2%) (one interviewee could mention more than one challenge). The success of any innovation will depend greatly on the users' awareness. The success of BECS, therefore, will depend on the awareness by its targeted users being the tourism operators. However, with most businesses still not aware of this tool, the success of the scheme, in the long run, remains questionable. Related to aspects of awareness, it is important to note that the interviewed tourism operators who did not think that tourists do consider eco-labels when selecting their tourism destinations argued that tourists usually consider the facility's prices, previous service satisfaction and the quality of the tourism facility and service over its eco-certification grade. These tourism operators thought that tourists would rather consider the already existing and known BTO's Star Rating System than the Eco-Certification Grading System when choosing their tourism destinations. One of the interviewees, therefore, concluded that there is no need to invest in the tourism certification scheme if tourists do not demand it yet. Hence one of the reasons for slow uptake could be attributed to inadequate pressure from tourists; hence eco-labelling awareness is not only important to operators but to tourists as well.

A small proportion of the respondents (15.8%) identified lack of skills as one of the main factors behind the slow uptake of the eco-certification scheme by tourism businesses. Already eco-certified operators were of the view that lack of expertise in the environmental sciences leads to most operations failing to comply with the strict requirements of the BECS. As noted above, some of the interviewees were Environmental Managers, a specialist post that was not available at other companies that were interviewed. Lack of staff specializing in environmental issues may therefore explain the reluctance to participate in the certification program by the companies. This is especially because BECS Accommodation Standards comprises more than 240 standards with the majority (52.9%) of them dealing with the minimization of negative impacts of the environment through physical design and operations. These standards require the knowledge of environmental experts who can perform the complex Environmental Management Systems with the relevant Environmental

Impact Assessments and Environmental Management Plans for their respective businesses.

Of the 63.2% of those operators who were aware of the BECS program, some of them indicated that they lacked understanding concerning the BECS program. They acknowledged that despite the BTO awareness campaigns that were conducted in the past, they still did not understand how the program operates hence their reluctance to apply for it. One manager stated that 'we always attend some of their (BTO) workshops and hear them saying 'go green, go green' but we don't really understand how it works. Since this statement shows that indeed efforts are being taken by BTO to raise awareness on certification, could it be then that lack of appropriate qualifications and hence the lack of required skills to incorporate the green practices in the operations of these businesses by those attending the workshops is still a challenge that hinders them from participating in the eco-labelling scheme? Hence Buckley (2002, p.185) opines that the response to any eco-label by stakeholders depends on factors such as: "the particular environmental issue or parameter to which the eco-label refers; the level of knowledge and concern among potential clients and other users in relation to environmental issues in general……; the degree of consensus regarding the meaning and significance of terms used in the eco-label".

Another factor that was identified by 26.3% of the interviewed businesses representatives was the lack of finances to support the needs of the BECS program and the financial burdens that come with it, such as enrolment fees and maintenance costs. For example, they decried high costs associated with setting up of solar energy systems, hiring environmental experts, establishments of recycling plants and waste disposal facilities which were needed to comply with the requirements of BECS. This may explain why only few big tourism operators could afford to enrol in the certification program. However, some proponents of tourism eco-labels argue that enrolment in these schemes would reduce the operating costs of the certified businesses in the long run as efficient businesses tend to attract more customers (Center Ecotourism and Sustainable Development, 2006). Therefore, companies should view this as an investment. Furthermore, these are the kind of initiatives that commercial banks which have adopted green financing could prioritize.

With respect to support tourism infrastructure, one of the interviewed companies which own several eco-certified businesses mentioned that the poor government infrastructure, such as the lack of a proper recycling plant, hindered their environmental performance as per the needs of the BECS program. It is worth noting that this problem persisted even after 5 years of having been identified by Mbaiwa et al. (2011). This may imply that the concerned departments being BTO, Waste Management and Pollution Control as well the District Councils, are working in silos instead of dovetailing their efforts. The interviewees were concerned that lack of supporting infrastructure might lead to some of their facilities failing to comply with the standards and their eco-certification status being revoked: "BTO is supposed to make regular checks to our camps, and if they find that you're not complying with the BECS standards, you risk your eco-certification grade being revoked which may have a negative publicity on your facility" said one manager.

According to the business representatives, waste management and carrying costs also increased, placing a negative burden on their financial position. This great challenge may discourage some operators, especially small businesses and Community Based Organizations that run tourism accommodation businesses, to enrol in the eco-certification program.

Of the interviewed respondents, 15.7% of businesses representatives mentioned that their business culture could be a factor that hinders them from enrolling in the BECS program. They mentioned that their business structures were set up under the traditional tourism regimes, that is, without any eco-friendly operations in their plans, and as a result, embracing new concepts to do with the role of tourism in sustainable development and the green economy was still a challenge. Those facilities that had some of their facilities certified mentioned that certification positively influenced their operations. One manager even mentioned that 'nowadays almost all of our business decisions are based on environmental considerations'. According to him, the certification has therefore led to a new business culture so much that awareness-raising for staff and clients on eco-friendly practices was also intensified to ensure that all relevant stakeholders contribute to the goal of keeping up with the certification standards.

3.6   Conclusion

Tourism is the second largest contributor to Botswana's Gross Domestic Product after mining, and most of the tourism activities take place in the Okavango and Chobe regions. However, as rightly pointed out by World Tourism Organization (2013), tourism can also be a source of environmental damage and pollution, a heavy consumer of scarce resources and a cause of negative impacts in society. For these reasons, it is imperative that it is well planned and managed, embracing the principles of sustainable tourism. According to Font (2002), efforts aimed at promoting sustainable tourism and ecotourism are disadvantaged by a lack of methods to ensure that the efforts do not become a part of 'green washing'. Various pathways that include accreditation schemes designed to operationalize one or more pillars of sustainable development have therefore been developed over time (Warnken et al., 2005). As observed, 'Eco-labelling is an instrument with potential to steer consumers as well as producers and the whole supply chains in a sustainable direction' (Bratt et al., 2011, p.1036). Efficient eco-labelling schemes should be characterized by social and environmental awareness, market dynamics, technological development, organizational strategy and environmental regulation and policy (Prieto-Sandoval et al., 2016). Therefore, eco-labelling provides cost-effective opportunities to induce more responsible environmental behaviours, given the environmental problems arising from unsustainable production and consumption (Minoli et al., 2015). This study, therefore, sought to unravel the reasons behind the seemingly slow involvement and acceptance of the BECS program by determining the perceptions of the Maun tourism accommodation services. As stated in previous studies, the motivation for participation is influenced by: altruism, self-interest, genuine concern for the environment, regarding the scheme as means of influencing future regulations, and consumer perception (Berghoef & Dodds, 2013). A study by Mbaiwa et al. (2011) investigated the prospects and challenges of tourism certification in Botswana, and this study comes 11 years later and adds to the contributions made by prior authors in this area. This study concludes that tourism businesses in Maun have positive perceptions about tourism eco-labelling and are willing to incorporate it into their businesses. However, several mitigation strategies such as raising awareness about BECS needs to be considered.

Even though this study did not cover the entire tourism sector in Botswana, it provides some important insights for policymakers. The insights from the study could be enhanced by further research seeking to determine the extent to which tourists demand and their willingness to pay for eco-certified tourism facilities as well as to assess how the eco-certified businesses have gained from implementing the program. For companies to be interested in participating in eco-labelling, they ought to understand and possibly be assured of the benefits since the process is expensive, and thereby participation should at least offset costs (Blackman et al., 2014).

References

Berghoef, N., & Dodds, R. (2013). Determinants of interest in eco-labelling in Ontario wine industry. *Journal of Cleaner Production, 52*, 263–271. https://doi.org/10.1016/j.jclepro.2013.02.020

Bernard, H. (1988). *Research methods in cultural anthropology*. Sage Publications.

Blackman, A., Naranjo, M. A., Robalino, J., Alpizar, F., & Rivera, J. (2014). Does tourism eco-certification pay? Costa Rica's Blue Flag Program. *World Development, 58*, 41–52. https://doi.org/10.1016/j.worlddev.2013.12.002

Botswana Tourism Board. (2002). Botswana ecotourism best practices manual. Retrieved from https://www.botswanatourism.co.bw/sites/default/files/publication/BTO%20EcoTourism%20Best%20Practices%20Manual.pdf

Botswana Trade and Investment Centre. (2016). Retrieved from: http://www.bitc.co.bw/tourism-licence

Bratt, C., Hallstedt, S., & Robert, K.-H. (2011). Assessment of eco-labelling criteria development from a strategic sustainability perspective. *Journal of Cleaner Production, 19*, 1631–1638. https://doi.org/10.1016/j.jclepro.2011.05.012

BTO (Botswana Tourism Organisation). (2010). *Botswana ecotourism certification system, accommodation standards*. Botswana Tourism Organisation.

Buckley, R. (2002). Tourism eco-labels. *Annals of Tourism Research, 29*(1), 183–208. https://doi.org/10.1016/S0160-7383(01)00035-4

Budeanu, A. (2005). Impacts and responsibilities for sustainable tourism: A tour operator's perspective. *Journal of Cleaner Production, 13*, 89–97. https://doi.org/10.1016/j.jclepro.2003.12.024

Center for Ecotourism and Sustainable Development. (2006). *A simple user's guide to certification or sustainable tourism and ecotourism*. Center for Ecotourism and Sustainable Development.

Chobe Game Lodge. (2016). Retrieved from http://www.chobegamelodge.co.bw/blog/chobe-game-lodge-launches-first-ever-eco-friendly-safaris/

Fair Trade Tourism. (2016). Retrieved from http://www.fairtrade.travel/blog/entry/fair-trade-tourism-joins-forces-with-botswanas-ecotourism-certification-sys

Font, X. (2002). Environmental certification in tourism and hospitality: Progress, process and prospects. *Tourism Management, 23*, 97–205. https://doi.org/10.1016/S0261-5177(01)00084-X

Font, X., & Buckley, R. C. (2001). *Tourism ecolabelling: Certification and promotion of & management.* CABI Publication.

Font, X., & Harris, C. (2004). Rethinking standards from green to sustainable. *Annals of Tourism Research, 31*(4), 986–1007. https://doi.org/10.1016/j.annals.2004.04.001

Githinji, W. M. (2006). *An evaluation of the use of ecolabelling within the ecotourism sector.* University of East Anglia.

Global Ecolabelling Network. (2016). Retrieved from http://www.globalecolabelling.net/what-is-ecolabelling/

Grand Palm Resort. (2016). Retrieved from http://www.grandpalm.co.za/Press-Room/Articles/grand-palm-eco-initiatives

Jensen, S., Birch, M., & Frederiksen, M. (2004). *Are tourists aware of tourism eco-labels? – results from a study in the county of Storstrøm in Denmark.* Paper presented at the 13th Nordic Symposium in Tourism and Hospitality Research, Aalborg, Denmark, November 4. -7, 2004, Aalborg University Tourism Research Unit, Aalborg.

Kitchin, R., & Tate, N. J. (2000). *Conducting research in human geography: Theory, methodology and practice.* Pearson Education Limited.

Kozak, M., & Nield, K. (2004). The role of quality and ecolabelling systems in destination benchmarking. *Journal of Sustainable Tourism, 12*(2), 138–148. https://doi.org/10.1080/09669580408667229

Lee, K. F. (2001). Sustainable tourism destinations: The importance of cleaner production. *Journal of Cleaner Production, 9*, 313–323.

Leonard, D. (2011). *A study into the motivation of participation in environemental certification by the tourism accommodation sector.* BA Thesis, MODUL University Vienna.

Mbaiwa, J. E. (2002). Environmental impacts of tourism development in the Okavango. *Journal of Arid Environments, 54*, 447–467.

Mbaiwa, J. E. (2015). Ecotourism in Botswana: 30 years later. *Journal of Ecotourism, 14*(2–3), 204–222. https://doi.org/10.1080/14724049.2015.1071378

Mbaiwa, J. E., Magole, L. I., & Kgathi, D. L. (2011). Prospects and challenges for tourism certification in Botswana. *Tourism Recreation Research, 36*(3), 259–270. https://doi.org/10.1080/02508281.2011.11081671

Minoli, D. M., Goode, M. M. H., & Smith, M. T. (2015). Are eco lables profitably employed in sustainable tourism? A case study on Audubon certified golf resorts. *Tourism Management Perspectives, 16*, 207–216. https://doi.org/10.1016/j.tmp.2015.07.011

Moswete, N., Mpotokwane, M. A., Nkape, K., & Maera, K. (2019). Exploring the potential and challenges: Guesthouse-based tourism and hospitality in Maun, Botswana. *African Journal of Hospitality, Tourism and Leisure, 8*(4). http://www.ajhtl.com/

Partington, G. (2001). Qualitative research interviews: Identifying problems in technique. *Issues in Educational Research, 11*(2), 32–44.

Petan, I. C., Silivia, D., & Murgoci, C. S. (2007). *Role of eco-labels in European Tourism Industry.* Faculty of Economic Sciences, University of Oradea.

Pieterse, G. H. (2004). Ecolabelling at lodges in South Africa. In F. D. Pineda & C. A. Brebbia (Eds.), *Sustainable tourism* (pp. 143–160). WIT Press.

Polkinghome, D. E. (2005). Language and meaning: Data collection in qualitative research. *Journal of Counselling Psychology, 52*(2), 137–145.

Prieto-Sandoval, V., Alfaro, J. A., Mejia-Villa, A., & Ormazabal, M. (2016). Eco-labels as a multidimenstional research topic: Trends and opportunities. *Journal of Cleaner Production, 135*, 806–818. https://doi.org/10.1016/j.jclepro.2016.06.167

Ruhanen, L., Weiler, B., Moyle, B. D., & McLennan, C. J. (2015). Trends and patterns in sustainable tourism research: A 25-year bibliometric analysis. *Journal of Sustainable Tourism, 23*(4), 517–535. https://doi.org/10.1080/09669582.2014.978790

Tepelus, C. M., & Cordoba, R. C. (2005). Recognition schemes in tourism-from 'eco' to 'sustainability'? *Journal of Cleaner Production, 13*, 135–140. https://doi.org/10.1016/j.jclepro.2003.12.015

Warnken, J., Bradley, M., & Guilding, C. (2005). Eco-resorts vs. mainstream accommodation providers: An investigation of the viability of benchmarking environmental performance. *Tourism Management, 26*, 367–379. https://doi.org/10.1016/j.tourman.2003.11.017

World Tourism Organization. (2013). *Sustainable tourism for development guidebook - enhancing capacities for sustainable tourism for development in developing countries*. UNWTO, Madrid.

Godiraone Trompies Motsaathebe is an aspiring sinologist (an expert in the study of China, Chinese language and culture), and scholar. He holds a Bachelor of Arts in Social Sciences from the University of Botswana and a certificate in Chinese Language Proficiency from the Confucius Institute at the University of Botswana.

Wame L. Hambira is a Senior Research Scholar with the Okavango Research Institute (University of Botswana). Her research interests include sustainable tourism and private sector sustainability practices. She holds a PhD in Geography from the University of Oulu, Finland and an MSc in Environmental Economics, University of York (UK).

Chapter 4
Inbound Tour Operator Participation in Sustainable Tourism Practices: A Focus on South Africa

Ignatius Ludolph Steyn, Felicite Fairer-Wessels, and Anneli Douglas

4.1 Introduction

Inbound tour operators fulfil a vital role in the tourism industry and play a critical role in sustainable tourism development. They are centrally positioned in the distribution chain, providing the link between supply and demand (Cavlek, 2002). Inbound tour operators can influence the direction of tourist flow, product offerings by tourism suppliers, and tourist attitudes and behaviour (Sigala, 2008). Many suppliers in the tourism industry also rely on tour operators to promote and distribute their products and services as they have limited resources. Embedded in this position, inbound tour operators can pressure their suppliers to operate more sustainably, educate tourists on sustainable tourism practices and influence consumer decision-making before purchasing tourism-related products and services. To date, little research has focussed on the contribution of inbound tour operators to sustainable tourism development, especially in a developing country context (Cavagnaro et al., 2015). Cavagnaro et al. (2015, p. 136) state that "… research on inbound tour operators (ITO) in general and in developing countries in particular is almost non-existent". One aspect contributing to sustainable tourism development is tour operators' membership in certification programmes.

Various studies have highlighted the history, benefits and issues related to certification programmes (CESD, 2007; Dodds & Joppe, 2005; Newton et al., 2004; Piper & Yeo, 2011). Studies have also been published in South Africa using Fair Trade Tourism as a case study (Boluk, 2011; Strambach & Surmeier, 2013). However, few studies have investigated inbound tour operators' perspectives

I. L. Steyn · F. Fairer-Wessels · A. Douglas (✉)
Tourism Management Division, Department of Marketing Management,
University of Pretoria, Pretoria, South Africa
e-mail: anneli.douglas@up.ac.za

47

J. Saarinen et al. (eds.), *Southern African Perspectives on Sustainable Tourism
Management*, Geographies of Tourism and Global Change,
https://doi.org/10.1007/978-3-030-99435-8_4

towards sustainable development tourism certification programmes. A study in South Africa investigated whether South African inbound tour operators put pressure on their supply-side to implement sustainable tourism practices and whether they experience pressure from their clients to implement sustainable tourism practices. The study also identified the value of sustainable tourism certification programmes in creating more sustainable South African inbound tour operators (Steyn, 2020).

The chapter consists of three parts. First, it reviews the existing literature concerning sustainable tourism, sustainable inbound tour operators, and sustainable tourism certification programmes. The chapter then explains how the specific study done in South Africa was approached and presents its findings to elucidate the current approaches to sustainable practices of inbound tour operators. The chapter concludes with an overview of the gaps in sustainability practices and proposes strategies that could produce more sustainable practices, particularly in South Africa.

4.2   Defining Sustainable Tourism

Sustainable tourism includes three main elements: environmental, socio-cultural, and economical. Together, these three elements are frequently defined as the 'triple bottom line'. Elkington (in CESD, 2007, p. 4) describes the triple bottom line as "... where a company examines the social, environmental and economic effects of its performance on the wider society, begins to improve its performance and reports publicly on progress". However, some research suggests that developed countries have devised the concept of sustainability, and it does not necessarily fit the needs of stakeholders in a developing country (Cavagnaro et al., 2015; Fox, 2004; Visser, 2008).

In developing countries, especially in Africa, poverty and unemployment are significant issues. Some researchers suggest that the socio-cultural element of the triple bottom line should take precedence over the environmental and economic elements (Butcher, 2011; Cavagnaro et al., 2015; Visser, 2008). The sustainability framework clearly states that there must be a balance between the ecological, socio-cultural, and economic dimensions of tourism development (Stoddard et al., 2012). Focussing only on the social dimension of sustainability would not promote this balance, as economic and environmental issues are also clearly apparent in developing countries. For this chapter, we use the UNWTO (2005) definition of sustainable tourism: "... tourism that takes full account of its current and future economic, social and environmental impacts, addressing the needs of visitors, the industry, the environment and host communities."

Sustainable and responsible tourism concepts have formed part of the development strategy of South Africa since 1996, reflected in the White Paper on the Development and Promotion of Tourism (DEAT, 1996) of the former Department of Environmental Affairs and Tourism (DEAT). In 2002, the DEAT further published Responsible Tourism Guidelines and a Responsible Tourism Manual to assist tourism organisations in understanding and implementing sustainable tourism practices (RTMSA, 2002; Goodwin et al., 2002). South Africa is currently the only African country that has developed National Minimum Standards for Responsible Tourism (NMSRT) (SANS 1162, 2011). Both the Department of Tourism and the former DEAT have shown that the concept of sustainable tourism is the preferred tourism development strategy for South Africa (DEAT, 1996). However, the Department of Tourism relies heavily on all South African tourism organisations to support and implement sustainable tourism practices.

4.3 Stakeholders in Sustainable Tourism Development

There are three stakeholders involved in sustainable tourism development: tourists/consumers, tourism business operators and governments. Consumers can influence the entire tourism industry by demanding sustainable tourism products. Although many research studies suggest that tourists are increasingly becoming more environmentally conscious (Barr et al., 2010; Bookings.com, 2019), recent studies have emphasised the difference in tourist willingness to travel sustainably and their actual purchasing behaviour (Anciaux, 2019; Budeanu, 2007; Miller et al., 2010). Tourism organisations are arguably in the best position to influence the industry to become more sustainable, as they are the suppliers offering tourism products and services. Tourism organisations appeared reluctant to make significant changes as the demand for sustainable tourism products remains a critical issue (Williams & Ponsford, 2009).

Regulators such as governments have the power to compel tourism organisations and consumers to implement sustainable tourism practices, enforcing them by law to cooperate. However, according to Williams and Ponsford (2009, p. 398), "…governments have little interest in burdening tourism businesses with additional regulations that might dampen their willingness to generate important tax revenues". Currently, these three stakeholders treat sustainable tourism as a 'hot potato', waiting for the other stakeholders to take leadership (Williams & Ponsford, 2009).

In addition, we can consider host communities as a stakeholder in sustainable tourism development. Besides direct employment created within the tourism industry through sustainable tourism development, supplying the industry with products and services can generate income. Studies suggest that this income is estimated to match, or even surpass, the income derived from direct employment within the tourism industry (Lengefeld & Stewart, 2004; Mitchell & Ashley, 2007). Thus, by including host communities in the decision-making and planning of sustainable tourism, various benefits may arise, such as the sharing of economic opportunities, the creation of additional entrepreneurial opportunities, educational opportunities, job creation, protection of environmental and cultural heritage and improved quality of life (Poudel et al., 2016). Therefore, it is crucial to investigate the sustainable tourism practices implemented by inbound tour operators. They can influence their demand and supply sides and positively impact the host communities visited.

4.4   Sustainable Tour Operator Development

A tour operator can be defined as an organisation that buys tourism-related products directly from the suppliers, combines them into attractive packages, and then sells them to customers (Budeanu, 2005). An inbound tour operator focuses primarily on bringing travellers into a country through group or individual tour packages (Saffery et al., 2007; Westcott et al., 2015).

In 2000 the Tour Operators' Initiative for Sustainable Tourism Development (TOI) was established. The TOI (2003, p. 10), supported by the United Nations Environmental Programme (UNEP), the United Nations World Tourism Organization (UNWTO) and the United Nations Educational, Scientific and Cultural Organization (UNESCO), aimed to "… improve the sustainability of the tourism industry, and to encourage tour operators to make a voluntary yet firm corporate commitment to sustainable development". Members of the TOI took action in three key areas: sustainability reporting, supply chain management, and cooperation with destinations.

4.4.1   Sustainability Reporting

In 2002 the TOI, in cooperation with the Global Reporting Initiative (GRI), developed the 'GRI Tour Operators' Sector Supplement', including it in the GRI 2002 Sustainability Reporting Guidelines. The supplement provides tour operators with a total of 47 performance indicators to help them measure and improve their sustainability performance (Dodds & Kuehnel, 2010). The GRI guidelines would assist mainly the more prominent and global tour operating companies in effectively producing a transparent sustainability report with measurable targets.

4.4.2   Supply Chain Management

Supply chain management, where the most significant impact can be made, was identified among the TOI members as a critical area requiring special attention leading to many tour operators developing and adopting a sustainable supply chain management strategy. Schwartz et al. (2008) state that a tour operator is only as sustainable as its suppliers.

Since tourists are also part of a tour operator's supply chain, tour operators can educate tourists, create awareness about the potential environmental and socio-cultural impacts that they may have on the destination, and provide information on avoiding adverse effects (Sigala, 2008).

4.4.3 Cooperation with Destinations

The third key area of action for TOI members highlights cooperation with destinations as, as TOI members "… work with a cross-section of stakeholders encompassing the diversity of views and interests present at the destination, including the local authorities, the private sector, civil society and NGOs" (TOI, 2003, p. 10).

In 2014 the TOI and the Global Sustainable Tourism Council (GSTC) decided to merge to jointly apply their energies and resources on work done through the GSTC's destination criteria, enabling entire destinations to become sustainably certified based on globally recognised standards (GSTC, 2018).

4.5 Sustainable Tourism Certification

4.5.1 Defining Certification

Honey and Rome (2001, p.8) define certification as "… a voluntary procedure that assesses, audits and gives written assurance that a facility, product, process or service meets specific standards. It awards a marketable logo to those that meet or exceed baseline standards". It is essential to distinguish between an eco-label and certification. Eco-labels imply that a product or service has considered its environmental impact (Piper & Yeo, 2011). Certification is instead a way of ensuring that an activity or a product meets specific standards (CESD, 2007).

4.5.2 Challenges and Developments Within the Certification

One of the biggest challenges concerning certification programmes, especially eco-labels, is that no regulation exists to prevent tourism organisations from self-declaring themselves as sustainable businesses. Many tourism organisations use their marketing platforms for advertising that their product offering helps conserve the environment or contribute to social and community aspects. However, these claims can be considered greenwashing without concrete evidence or affiliation with a credible association or certification programme. Greenwashing refers to a business claiming to be 'sustainable', 'eco' or 'green', when it does not comply with a generally accepted standard, or worse, contradicts them (CESD, 2007).

Aware of this issue, a study conducted by the Sustainable Tourism Stewardship Council (STSC) in 2002 recommended forming an accreditation body to certify the certifiers (Dodds & Joppe, 2005). In 2007, the 'Partnership for Global Sustainable Tourism Criteria' (GSTC) was formed as a coalition of 32 partners, initiated by the Rainforest Alliance, the UNEP, the United Nations Foundation (UNF) and the UNWTO. In 2008, the GSTC Partnership developed a set of baseline criteria organised around the four pillars of sustainable tourism: "… effective sustainability planning; maximising social and economic benefits to the local community; reduction of negative impacts to cultural heritage; and reduction of negative impacts to the environment" (GSTC, 2018). Today the GSTC's certification criteria serve as the global baseline standard for sustainability in tourism and travel.

It is important to note that certification is not a substitute for sound business practices. Various authors state that sustainability only becomes part of a consumer's decision-making once the primary criteria of price, quality, safety, availability, and location are met (Dodds & Joppe, 2005; Tasci, 2017). However, accommodation suppliers implementing sustainable tourism practices have shown that in the long-term, sustainability can increase the quality of their product and service, offer customer satisfaction, and also lower the prices for many products and services (Tasci, 2017). One example is TUI (Touristik Union International). Once TUI realised that their most sustainably managed hotels deliver higher quality and customer satisfaction, they started to encourage all their hotels to become certified by a credible certification programme. TUI's hotel partners currently have a mandatory clause in their contracts, requiring them to work towards a certification recognised by the GSTC (TUI, 2019).

The second issue regarding certification is the lack of consumer demand for sustainable tourism products and services. Research studies have found a significant disparity between the increasing claims by consumers to be environmentally aware and their actual purchasing behaviour (Baddeley & Font, 2011; Dodds & Kuehnel, 2010; Font & Wood, 2007; Miller et al., 2010). Research conducted by Booking.com, one of the world's leading online travel agencies, found that 70% of global travellers "… would be more likely to book an accommodation knowing it was eco-friendly, whether they were looking for a sustainable stay or not. Well over a third (37%) of [travellers] affirm that an international standard for identifying eco-friendly accommodation would help to encourage them to travel more sustainably" (Booking.com, 2019).

However, in 2017 Tasci asked 411 American residents to provide any three performance measures used within the tourism and hospitality industry, including benchmarks, standards, licences, or certification. Fifty per cent (50%) of the respondents were not able to provide any name, but more importantly, none mentioned the major 'globally recognised' sustainable tourism certification programmes. Tasci (2017) suggests that certification programmes should instead adopt a marketing approach because tourism organisations will become more reluctant to become certified if consumers are unaware of their names. However, certification programmes are run mainly by government agencies or NGOs lacking the necessary marketing experience and finances to promote certified tourism organisations effectively. They are also reluctant to get involved with marketing companies as it conflicts with their core business, evaluating standards (Font & Wood, 2007).

Ultimately, tour operators and wholesalers, considered 'key buyers', do not actively promote certified suppliers. This is because "… (i) the labels are not consistently available around the world, and (ii) they do not represent meaningful value to the customer" (Font & Wood, 2007, p. 158).

4.5.3   Certification in South Africa

In South Africa, a few sustainable tourism certification programmes exist, including the Green Leaf Eco Standard (GLES), Greenline, and Fair Trade Tourism (FTT) (FTT, 2018; GLES, 2018; GreenLine, 2018). FTT is the only certification programme in South Africa recognised by the GSTC and the only tourism organisation globally that incorporates fair trade principles into its certification criteria. Although not considered a certification but rather eco-labelling, FTT does 'approve' tour operators and award them with a marketable logo in exchange for their help to support and promote FTT certified accommodation suppliers, thereby encouraging a sustainable supply chain. FTT approved tour operators must have at least one packaged holiday or itinerary with 50% of their bed nights at FTT certified or mutually recognised accommodation suppliers (FTT, 2018). Only the international and GSTC accredited certification programme, Travelife for Tour Operators, certifies tour operators in South Africa. However, only three tour operators have been approved (Travelife, 2019).

4.6   Case Study: South Africa

4.6.1   Background

In 2020, Steyn (2020) conducted a study to investigate the sustainable tourism practices in South African inbound tour operators. In-depth interviews were conducted with 22 South African inbound tour operators to examine whether they feel pressure from their demand-side to operate more sustainably and whether they put pressure on their supply-side to implement sustainable tourism practices. Also, this study aimed to identify the value that sustainable tourism certification programmes add to the development of more sustainable inbound tour operators (Steyn, 2020).

The final sample consisted of 22 South African inbound tour operators, of which 11 were 'approved' or certified by a sustainable tourism certification programme, and 11 neither certified nor 'approved', but are actively operating in the industry.

4.6.2   Sustainable Supply Chain Management (Supply-Side)

The results of the study by Steyn (2020) confirmed the findings of Spenceley (2006), Van der Merwe and Wöcke (2007) and Cavagnaro et al. (2015) that in the African tourism industry, the social element of the triple bottom line had received slightly more attention than the environmental element.

> I think in South Africa we do not realise that we are in a unique situation … What I find almost always, if it is ecotourism it includes local society and people, and taking care of them.

Similarly, most participants felt there is a ripple effect through tourism. By simply bringing international clients to South Africa, the country will benefit economically. Still, this mindset can seriously threaten sustainable tourism development as the tourism industry is criticised for destroying the natural and cultural resources upon which the destination is based (Dlamini & Masuku, 2013; Poudel et al., 2016).

The findings did reveal that participants take their prospective suppliers' sustainable tourism practices into consideration when developing travel itineraries. These findings support the result of Feruzi et al. (2013), which states that 80.8% of their respondents, being tour operators in Tanzania, agreed to support accommodation providers who follow specific ecotourism principles. This preference towards sustainable suppliers may result from the host destination and the expectations of tourists when visiting an African country. Inbound tour operators prefer accommodations suppliers with social and environmental projects to add a unique element of sustainability to their clients' visit, ensuring a memorable 'African' experience. It is important to note that sustainability is not the leading factor when selecting products and services. According to the study participants, the most significant influence on decisions is requests received from their clients. Preference is only given to sustainable suppliers if they provide the required location and price, as suggested by Dodds and Joppe (2005) and Tasci (2017):

> Sometimes we can suggest what we think is best for the client and we can then choose what we like, but often times the client has done research and knows where they want to go.

Seven participants confirmed that they put pressure on their accommodation suppliers to become more sustainable. Eight participants said their organisations were reluctant to pressure accommodation suppliers. They instead encouraged them to become more sustainable by providing feedback on practices they or their clients discourage or find unsustainable: *"I have found that when you pressurise people, you get more resistance and they become more adamant. I do not think pressure is the best way to do it. I think educating is better."*

Findings further revealed that some smaller organisations felt they had no bargaining power over accommodation suppliers and could not influence or pressure their suppliers to operate more sustainably. Of concern is that some of the larger organisations with little bargaining power over their suppliers stated that it is not their responsibility or duty to put pressure on suppliers to implement sustainable tourism practices: *"We are in a position to put pressure on them. It is a huge job, but it is not our job. I think they themselves should become more responsible."*

4.6.3   Sustainable Supply Chain Management (Demand-Side)

None of the participants indicated that their organisations felt pressure from their demand-side to implement sustainable tourism practices. Neither their clients nor agents asked about the organisation's sustainable tourism practices before purchasing their products or services: *"It is one of the few sectors where the change has been pushed by the industry rather than on demand from the clients"*.

Findings from the study showed that tourists' demand for sustainable tourism products and services is growing. The study also found that, as tourists arrive in South Africa, their environmental, social, and economic awareness increases: *"I do not think it is at a point where a tourist will say we won't stay anywhere that is not sustainable or supporting sustainable tourism, but there is a greater awareness. When people get here, they ask more questions, and they ask more questions of the lodges."*

Participating organisations agreed with Orgams (in Tasci, 2017) that education is vital for sustainable tourism development while at the same time contributing towards overall quality and customer satisfaction. The majority of the participants indicated that they are the ones that would foster demand by educating their clients on how to travel sustainably. The organisations' tour guides inform their clients about environmental or social issues before visiting a host community, as do their accommodation suppliers running conservation and community-related projects. Other educational practices mentioned are providing clients with responsible travel tips, highlighting the organisation's conservation and community projects before, during and after the tour, and educating clients through marketing and promotional material.

The tour guide might arrive at a specific site and tell them what interesting things they have done at the place, like how they recycle water. We educate them in the townships. There is a lot of education on how the communities there create vegetable gardens ...

4.6.4 The Value of Certification Programmes in Sustainable Tourism Development

Participants from both certified and non-certified organisations revealed that they do find value in sustainable tourism certification programmes. The value that sustainable tourism certification programmes add to the development of sustainable inbound tour operators has been identified as the following:

- Information sharing, especially the sharing of best practices.
- Providing guidance and recommendations on how to approach sustainability issues.
- Creating general awareness related to sustainable tourism.
- Providing opportunities to stakeholders to network with like-minded individuals.
- Providing sustainability training to members.
- Certifying inbound tour operators based on national or globally recognised standards.
- Assisting in promoting and branding sustainably certified inbound tour operators by providing a marketable logo.

- Justifying, through certification standards and periodic audits, that an inbound tour operator is truly operating sustainably and that they are not subject to greenwashing.
- Assisting inbound tour operators in identifying truly sustainable and certified accommodation and activity/excursion suppliers and destinations to aid in developing a truly sustainable supply chain and minimise the effects of greenwashing

Although various value-adding elements have been identified, very few organisations participate in sustainable tourism certification in the broader scope of the South African tourism industry. Only three tour operators in South Africa are currently certified by a sustainable tourism certification programme. The low uptake in certification may be directly related to the low levels of awareness and demand for certified products and services, as suggested by Tasci (2017) and Jarvis et al. (2010). Font and Wood (2007) state that both lodge owners and tour operators find no desire from their clients to meet "green standards". Furthermore, membership of a certification programme was perceived to provide little benefit for participating organisations, suggesting that certification programmes should promote/market and educate both end-users and tourism suppliers, as indicated by Tasci (2017). This sentiment was reflected in comments such as: *"I think that they should be doing a lot of the education … They should have two arms for their education programmes. They should be educating the operators and industry professionals, but they should also be educating the end-user, the consumers."*

Participants from both certified and non-certified/approved organisations highlighted the issue of greenwashing. Some of the non-certified/approved organisations felt that without being audited against a set standard, no organisation should be allowed to receive any form of eco-label. Participants from certified organisations felt that certification protects them from greenwashing, showing their clients, suppliers, and industry commitment to sustainability. They are being audited against set standards every 3 years. According to Font and Wood (2007) certification can help protect customers against false claims: *"There is such an enormous amount of greenwashing going around, it makes it very tangible and makes us accountable to our clients because we are independently audited. It shows our clients that we do not just talk the talk but walk the walk."*

4.7   Gaps and Recommendations

This chapter highlighted issues that need to be addressed by all stakeholders involved in the South African tourism industry, both private and governmental. South African inbound tour operators need to actively promote and market sustainable tourism products and services and educate their clients on the concept of sustainable tourism to drive and foster demand. Tour operators should actively support sustainable tourism suppliers giving preference to certified suppliers of an accredited sustainable tourism certification programme to encourage non-sustainable suppliers to operate more sustainably.

South African tourism suppliers should use the DEAT's Responsible Tourism Guidelines and the Responsible Tourism Manual to educate themselves on implementing sustainable tourism practices cost-effectively. Organisations may use the NMSRT as a tool when developing their organisation's sustainable tourism strategy, goals and objectives to work towards the national benchmark. Alternatively, South African tourism suppliers can become members of a sustainable tourism certification programme to receive training on how to operate sustainably and start the process towards becoming a truly sustainable tourism organisation free from greenwashing.

The government needs to drive education, create awareness of the importance of sustainable tourism, and create awareness of its published documentation related to responsible/sustainable tourism. The government should assist and promote sustainable tourism certification programmes to become accredited by SANAS, the South African National Accreditation System, and certify tourism organisations against the NMSRT. Currently, no certification programme is accredited to use these standards (SANAS, 2019).

South African certification programmes should consider certifying with two levels of standards. The first level should align with the NMSRT, with the second aligning its higher-level criteria with GSTC's standards. Various participants mentioned that the current standards for FTT and Travelife are too high and almost impossible to reach. On a national and a global level, two levels of standards may encourage tourism organisations to start small and build themselves toward a globally recognised standard. Ultimately, all stakeholders in the South African tourism industry will need to work together and align their strategies to achieve the primary goal of sustainability.

Not all income generated through inbound tourism is directly related to inbound tour operators. Other booking methods can be utilised, such as direct bookings and bookings via online travel agencies (OTAs). Further research may explore possible partnerships and collaborations between GSTC accredited certification programmes and large OTAs such as Booking Holdings and the Expedia Group to promote sustainable and certified tourism suppliers, create awareness among tourists, and foster demand for sustainably accredited tourism products and services.

4.8 Conclusion

This chapter explained the role of South African inbound tour operators in sustainable tourism development. The chapter covered how clients influence inbound tour operators to operate more sustainably by placing pressure on their suppliers to manage more sustainably, given their central position between the supply and demand for tourism products and services. The chapter also discussed the value of sustainable tourism certification programmes in developing more sustainable South African inbound tour operators.

This chapter used the findings of an academic study conducted in South Africa that revealed that although no pressure is received from consumers for sustainable tourism products and services, South African inbound tour operators prefer working with sustainable suppliers to assist in creating a memorable 'African' experience. This study by Steyn (2020) proposed that sustainable tourism organisations should become certified by a national or global sustainable tourism certification programme to prove that they are genuinely operating sustainably, thus decreasing the effects of greenwashing. In addition, the certification of tourism organisations can assist inbound tour operators in identifying truly sustainable suppliers, thereby aiding the development of a sustainable supply chain management strategy.

References

Anciaux, A. (2019). "On holidays, I forget everything... Even my ecological footprint": Sustainable tourism through daily practices or compartmentalisation as a keyword? *Sustainability, 11*(4731), 1–19.

Baddeley, J., & Font, X. (2011). Barriers to tour operator sustainable supply chain management. *Tourism Recreation Research, 36*(3), 205–214.

Barr, S., Shaw, G., Coles, T., & Prillwitz, J. (2010). "A holiday is a holiday": Practicing sustainability, home and away. *Journal of Transport Geography, 18*(3), 474–481.

Boluk, K. (2011). Fair trade tourism South Africa: Consumer virtue or moral selving? *Journal of Ecotourism, 10*(3), 235–249.

Bookings.com. (2019). *Booking.com reveals key findings from its 2019 sustainable travel report.* Retrieved June 11, 2019, from https://globalnews.booking.com/bookingcom-reveals-key-findings-from-its-2019-sustainable-travel-report

Budeanu, A. (2005). Impacts and responsibilities for sustainable tourism: A tour operator's perspective. *Journal of Cleaner Production, 13*, 89–97.

Budeanu, A. (2007). Sustainable tourist behaviour - a discussion of opportunities for change. *International Journal of Consumer Studies, 31*, 499–508.

Butcher, J. (2011). Can ecotourism contribute to tackling poverty? The importance of "symbiosis". *Current Issues in Tourism, 14*(3), 295–307.

Cavagnaro, E., Staffieri, S., & Ngesa, F. (2015). Looking from a local lens: Inbound tour operators and sustainable tourism in Kenya. *Research in Hospitality Management, 5*(2), 135–145.

Cavlek, N. (2002). Tour operators and sustainable development - a contribution to the environment. *Journal of Transnational Management Development, 7*(4), 45–54.

CESD. (2007). *A simple user's guide to certification for sustainable tourism and ecotourism* (3rd ed.). A publication of the Center for Ecotourism and Sustainable Development.

DEAT, Department of Environmental Affairs and Tourism. (1996). *White Paper on the Development and promotion of tourism in South Africa.* Government of South Africa.

Dlamini, C., & Masuku, M. (2013). Towards sustainable financing of protected areas: A brief overview of pertinent issues. *International Journal of Biodiversity and Conservation, 5*(8), 436–445.

Dodds, R., & Joppe, M. (2005). *CSR in the tourism industry? The Status of and potential of certification, codes of conduct and guidelines.* Study prepared for the CSR practice Ffreign investment advisory service investment climate department.

Dodds, R., & Kuehnel, J. (2010). CSR among Canadian mass tour operators: Good awareness but little action. *International Journal of Contemporary Hospitality Management, 22*(2), 221–244.

Feruzi, J., Steyn, J., & Reynisch, N. (2013). The extent to which Tanzanian tour operators apply sustainable practices as outlined in the national tourism policy. *African Journal for Physical, Health Education, Recreation and Dance, 19*(1), 82–96.

Font, X., & Wood, M. (2007). Sustainable tourism certification marketing and its contribution to SME market access. In R. Black & A. Crabtree (Eds.), *Quality assurance and certification in ecotourism* (pp. 147–163). CAB International.

Fox, T. (2004). Corporate social responsibility in development: In quest of an agenda. *Development, 47*(3), 29–36.

FTT. (2018). *Fair trade tourism*. Retrieved March 27, 2018, from http://www.fairtrade.travel/Home/

GLES, Green Leaf Eco Standard. (2018). *The Standard*. Retrieved June 08, 2018, from http://www.greenleafecostandard.net/the-standard.html

Goodwin, H., Spenceley, A., & Maynard, B. (2002). *Development of responsible tourism guidelines*. Department of Environmental Affairs and Tourism.

GreenLine. (2018). *Recognition*. Retrieved June 08, 2018, from http://www.greenline-rt.com/recognition

GSTC. (2018). *Global sustainable tourism council*. Retrieved March 22, 2018, from https://www.gstcouncil.org/about/about-us/

Honey, M., & Rome, A. (2001). *Protecting paradise: Certification programs for sustainable tourism and ecotourism*. Institute for Policy Studies.

Jarvis, N., Weeden, C., & Simcock, N. (2010). The benefits and challenges of sustainable tourism certification: A case study of the green tourism business scheme in the west of England. *Journal of Hospitality and Tourism Management, 17*(1), 83–93.

Lengefeld, K., & Stewart, R. (2004). *All-inclusive resorts and local development: Sandals as best practice in the Caribbean*. Presentation at the WTM in London.

Miller, G., Rathouse, K., Scarles, C., Holmes, K., & Tribe, J. (2010). Public understanding of sustainable tourism. *Annals of Tourism Research, 37*(3), 627–645.

Mitchell, J., & Ashley, C. (2007). Can tourism offer pro-poor pathways to prosperity? Examining evidence on the impact of tourism on poverty. *ODI Briefing Paper, 22*(1), 4.

Newton, T., Quiros, N., Crimmins, A., Blodgett, A., Kapur, K., Lin, H., … Dunivan, D. (2004). *Assessing the certification of sustainable tourism program in Costa Rica, School for Field Studies*. Center for Sustainable Development Studies.

Piper, L., & Yeo, M. (2011). Ecolabels, ecocertification and ecotourism. In *Sustainable tourism: Socio-cultural, environmental and economics impact* (pp. 279–294). University of Rijeka.

Poudel, S., Nyaupane, G., & Budruk, M. (2016). Stakeholders' perspectives of sustainable tourism development: A new approach to measuring outcomes. *Journal of Travel Research, 55*(4), 465–480.

RTMSA. (2002). *Responsible tourism manual for South Africa*. Department of Environmental Affairs and Tourism.

Saffery, A., Morgan, M., Tulga, O., & Warren, T. (2007). *The business of inbound tour operators: Tour operators manual*. United States Agency for International Development.

SANAS. (2019). *South African National Accreditation System*. Retrieved June 11, 2019, from https://www.sanas.co.za/

SANS 1162. (2011). *National minimum standard for responsible (NMSRT)*. SABS Standards Division.

Schwartz, K., Trapper, R., & Font, X. (2008). A sustainable supply chain management framework for tour operators. *Journal of Sustainable Tourism, 16*(3), 298–314.

Sigala, M. (2008). A supply chain management approach for investigating the role of tour operators on sustainable tourism: The case of TUI. *Journal of Cleaner Production, 16*, 1589–1599.

Spenceley, A. (2006). *Responsible tourism practices by south African tour operators*. International Centre for Responsible Tourism - South Africa.

Steyn, I. (2020). *Investigating south African inbound tour operator participation in sustainable tourism practices*. Master's dissertation, Department of Marketing, Division: Tourism Management, University of Pretoria.

Stoddard, J., Pollard, C., & Evans, M. (2012). The triple bottom line: A framework for sustainable tourism development. *International Journal of Hospitality & Tourism Administration, 13*(3), 233–258.

Strambach, S., & Surmeier, A. (2013). Knowledge dynamics in setting sustainable standards in tourism–the case of 'Fair trade in tourism South Africa'. *Current Issues in Tourism, 16*(7–8), 736–752.

Tasci, A. (2017). Consumer demand for sustainability benchmarks in tourism and hospitality. *Tourism Review, 72*(4), 375–391.

TOI (Tour Operator Initiative). (2003). *Sustainable tourism: The tour operators' contribution*. TOI.

Travelife. (2019). *Travelife for tour operators and travel agencies*. Retrieved June 11, 2019, from https://www.travelife.info/

TUI (Touristik Union International). (2019). *Responsibility*. Retrieved June 13, 2019, from https://www.tuigroup.com/en-en/responsibility/sus_business/hotel

UNWTO. (2005). World tourism organization, sustainable tourism for development. [Online] Available at: http://genevaoffice.unwto.org/content/about-us-5 [Accessed 01 06 2018].

Van der Merwe, M., & Wöcke, A. (2007). An investigation into responsible tourism practices in the south African hotel industry. *South African Journal of Business Management, 38*(2), 1–15.

Visser, W. (2008). Corporate social responsibility in developing countries. In A. Crane, A. McWilliams, D. Matten, J. Moon, & D. Sgel (Eds.), *The Oxford handbook of CSR* (pp. 473–479). Oxford University Press.

Westcott, M., Webster, D., Owens, D., Thomlinson, E., Bird, G., Tripp, G., ... Hood, T. (2015). In M. Westcott (Ed.), *Introduction to tourism and hospitality in BC*. BCcampus.

Williams, P., & Ponsford, I. (2009). Confronting tourism's environmental paradox: Transitioning for sustainable tourism. *Futures, 41*, 396–404.

Ignatius Ludolph Steyn is a Master of Commerce in Tourism Management, University of Pretoria, South Africa. His research interests are sustainable tourism management, sustainable supply chain management in tourism, sustainable tourism certification programmes.

Felicite Fairer-Wessels is an Emerita Professor at the University of Pretoria, Department of Marketing Management, Division Tourism Management. Her research interests are sustainable tourism development, sustainable development goals, climate change, cultural and heritage tourism, volunteer tourism, tourism information exchange, attractions and events.

Anneli Douglas is an Associate Professor in the Division: Tourism Management at the University of Pretoria, Pretoria, South Africa. Her research interests are focusing on business and corporate travel management, nature-based tourism experiences.

Chapter 5
Tourism-Led Inclusive Growth Paradigm: Opportunities and Challenges in the Agricultural Food Supply Chain in Livingstone, Zambia

Brenda M. K. Nsanzya and Jarkko Saarinen

5.1 Introduction

In many low and middle-income countries, the tourism industry has been considered as a favourable tool for development and economic diversification (Anderson, 2018; Bakker & Messerli, 2017). Particularly in emerging economies in Africa, the growth of tourism has been accompanied by rising GDP, foreign exchange, infrastructural development, employment creation, enterprises opportunities and infrastructure expansion (see Lacher & Nepal, 2010; Torres & Momsen, 2004; Saarinen, 2020). However, inequality and poverty persist, and tourism's potential contribution to the Sustainable Development Goals (SDGs) and related targets, such as reducing poverty, creating inclusive economic linkages and improvement of living standards, remain largely unknown or underdeveloped (Rogerson & Rogerson, 2020; Scheyvens et al., 2021; Spencer et al., 2014). This has created a growing academic interest in investigating real net benefits of tourism among destination communities (Jeyacheya & Hampton, 2020; Mbaiwa, 2005; Meyer, 2012; Rogerson, 2013; Saarinen, 2019).

Recently, within inclusive growth thinking, which is recognised as one of the United Nation's Sustainable Development Goals (SDG 8), debates on connections and disconnections between tourism and development have gained increasing

B. M. K. Nsanzya (✉)
Tourism Management Division, Department of Marketing Management, University of Pretoria, Hatfield, South Africa
e-mail: brenda.nsanzya@up.ac.za

J. Saarinen
Geography Research Unit, University of Oulu, Oulu, Finland

School of Tourism and Hospitality, University of Johannesburg, Johannesburg, South Africa
e-mail: jarkko.saarinen@oulu.fi

© The Author(s), under exclusive license to Springer Nature   61
Switzerland AG 2022
J. Saarinen et al. (eds.), *Southern African Perspectives on Sustainable Tourism Management*, Geographies of Tourism and Global Change,
https://doi.org/10.1007/978-3-030-99435-8_5

attention among tourism scholars (Bakker & Messerli, 2017; Biddulph & Scheyvens, 2018; Saarinen, 2017, 2020). Jeyacheya and Hampton (2020) have stated that despite tourism's potential for strengthening linkages to food and non-food sectors, for example, the tourism-led inclusive growth paradigm may widen inequalities in host destinations, weaken backward linkages to the local economy and limit opportunities to economic growth and social development (Schilcher, 2007). Therefore, opportunities and challenges exist within the tourism-led inclusive growth paradigm (Biddulph & Scheyvens, 2018; Saarinen & Rogerson, 2014).

This highlights the need for a greater assessment of tourism-led growth at the micro-level as experienced by local communities. This chapter focuses on the tourism-led inclusive growth paradigm and opportunities and challenges that it may create in the agricultural food supply chain. While linkages between tourism and other economic sectors like agriculture are possible, there are a number of challenging factors that need to be identified and mitigated as these affect the successful implementation and sustainability of market linkages (Meyer, 2007; Torres & Momsen, 2011), which have a bearing on tourism-led inclusive growth (Biddulph & Scheyvens, 2018; Jeyacheya & Hampton, 2020). Using evidence from empirical research in Livingstone, Zambia, this paper contributes to the emerging debate on inclusive growth, specifically by evaluating challenges and opportunities within tourism-agriculture linkages. Within the pro-poor tourism theoretical framework, the paper highlights facilitators and barriers to sustainable tourism-agriculture market linkage.

5.2   Inclusive Growth: Tourism – Agriculture Nexus

In the pre-COVID-19 economy, global tourism was on a consistent growth path (UNWTO, 2020). Although there are uncertainties, it is expected that both domestic and international tourism will return to a growth path in the post-COVID economy (Prideaux et al., 2020). In normal circumstances, tourism is one of the sectors that can broaden the distribution of income and wealth through productive employment, businesses, and other opportunities, including economic linkages (Anderson, 2018; Jeyacheya & Hampton, 2020). As a place-based industry highly dependent on its geographical surroundings (Hall & Lew, 2009) and other economic sectors for goods and services, partnerships are an important mechanism for a sustained tourism economy, creating opportunities for linkages between the tourism industry and non-tourism sectors in a tourist destination (Hall & Page, 2014; Pillay & Rogerson, 2013). The possibility of developing alliances is further supported by Van der Duim, Meyer and Saarinen (2011, p. 17) in their assertion that within the growing tourism industry, it is possible to develop "new alliances" between the tourism industry and other economic sectors in tourism destination environments.

These new alliances and linkages call for a symbiotic relationship that is mutually beneficial, economically viable and preferably inclusive (Oyinlola et al., 2020). Not only do linkages foster revenue retention and circulation within a locale, but they are an important aspect of tourism development as they may create opportunities for the reduction of economic leakages that are typical, especially in low and middle-income countries (Kavita & Saarinen, 2016; Mbaiwa, 2005). This, in turn, stimulates local economic development in the destination economy and creates synergetic effects between the various sectors (Meyer, 2012; Uduji et al., 2020) and sub-sectors such as tour operators, hotels, guest houses, food wholesalers, transport, farmers, construction, crafts and souvenir shops (Anderson, 2018; Mitchell, 2010a, b). In this respect, there are several factors that create the synergy between tourism and agriculture and also determine their synergetic relationship in sustainability. Their relationship is based on a division of labour: a supply and demand in local and regional economies (Pillay & Rogerson, 2013). In addition, the characteristics of this relationship may threaten or enrich the quality of the tourism-agriculture linkages. These characteristics relate to supply or production, demand, the role of government and policy and the role of intermediaries and marketing (Meyer, 2007; Meyer et al., 2004; Rylance et al., 2009; Torres & Momsen, 2004).

In the context of linking tourism to agriculture, Berno (2011, p. 87) puts it simply that "better linkages between agriculture and tourism contribute to the ethos of sustainable tourism". She further emphasises that greater linkages between the two sectors can result in higher levels of economic retention and contribute significantly to sustainable tourism. The Food and Agriculture Organisation of the United Nations (FAO) has advanced this view by cautioning that the synergy between the two sectors ought to yield outcomes that mutually reinforce each sector as opposed to creating competition for productive resources (FAO, 2012). For example, local sourcing of food products reduces costs and improves the quality of products, as a result significantly fostering the responsibility and social licence for tourism businesses to operate (World Bank, 2012). This assertion points to the potential for the global tourism industry to be inclusive as a sector operating in local or national economies that are expanding and thriving through an enhancement of tourism-agriculture linkages.

There is evidence of growing interest among tourism scholars investigating tourism-agriculture linkage (Rogerson & Rogerson, 2020; Uduji et al., 2020). A few examples include empirical research from the African continent, which show contrasting views on the practice of the linkages between the two sectors. For example, strong linkages were found in the case of Lushoto, Northeast Tanzania, where tourism value chains stimulate the economy (Anderson, 2018). In other cases, tourism-agriculture linkages exist but are not necessarily based on a pro-poor approach, as demonstrated in South Africa by Rogerson (2013). Weak linkages between tourism-agriculture were also reported in Botswana (Hunt et al., 2012) and Southern Ethiopia (Bale Mountains National Park), whereby the synergy between the two economic sectors presented no economically profitable coexistence between them (Welteji & Zerihun, 2018).

While there is growing interest in inclusive growth, in general, and specific sector-led inclusive growth, there is still limited research on tourism-linked inclusive growth (Bakker & Messerli, 2017; Rogerson & Rogerson, 2020). Hampton et al. (2017, p. 359) stated that very few studies have empirically tested the inclusive growth notion within the tourism sector. However, the general perception that tourism automatically drives economic development in emerging economies has been partly challenged, and especially the assumption that tourism revenues trickle down to other sectors and communities is increasingly questioned (Hunt et al., 2012; Mbaiwa, 2005; Saarinen, 2016). Based on this, the tourism industry's contribution to small-scale enterprises in the agriculture sector and stakeholders in the tourism supply chain requires thorough assessments. Indeed, Torres and Momsen (2004) have articulated the significance of research investigating the link between tourism and agriculture as "necessary to achieving the pro-poor dual objectives of reducing negative impacts while generating net benefits for the poor" (2004, p. 299). This assertion is reflective of an inclusive rather than exclusive tourism-agriculture linkage.

5.3 Tourism-Led Inclusive Growth in Livingstone, Zambia

5.3.1 Case Site and Methodology

Located in the southern part of Zambia, Livingstone is 10 km north of the Zambezi River and Victoria Falls, Africa's highest waterfall. The Victoria Falls is located by the Mosi-oa-Tunya National Park and forms part of Livingstone District's border with Zimbabwe. Victoria Falls was designated a UNESCO World Heritage Site in 1989, and it is regarded as one of the 'Seven Wonders of the Natural World' (McLachlan & Binns, 2014). It is a key tourism attraction of Zambia, drawing greater numbers of visitors than all the five major national parks in the country combined.

Tourism has been the driver of Livingstone's economy. It is endowed with adventure-based tourism activities, wildlife, the Zambezi River and the Victoria Falls, attracting international tourists (McLachlan & Binns, 2014). This is a significant shift from the early 1990s when Livingstone was considered a "ghost town". This change resulted from tourism-related investments and developments from the early 2000s when the state embarked on promoting Zambia as a tourist destination to tourists and investors as well as the associated infrastructure development. This included the expansion and renovation of the Livingstone Airport, which was upgraded to a national airport and completed in 2017. A Ministry of Tourism and Arts (2017, p. 3) report shows that in 2017 there were 178,714 tourist arrivals in the region; an increase of 14,272 (7.9%) from 2016's recorded 164,442 tourist arrivals. The upward increase in international tourist arrivals is consistent with global trends.

There are 130 accommodation enterprises registered and operating in the Livingstone and surrounding Kazungula district. The range of accommodation establishments includes 11 hotels, 72 lodges, over 20 guest houses and some accommodation establishments registered as inns, campsites, and apartments. With a growing tourism industry, accommodation establishments in Livingstone and surrounding areas are an important source of investment by local residents, creating employment and income-generating opportunities. Sustainable local procurement by the tourism sector presents an opportunity for communities in the periphery of the local economy, such as small-holder farmers, to participate and benefit from the tourism economy (Ministry of Tourism and Arts, 2017).

While the 'sustainable' tourism and 'inter-sectoral linkages' rhetoric is evident in the current national tourism policy (Ministry of Tourism, 2015), it is open how sustainable and economically viable tourism-agriculture linkages are, particularly in and around the Livingstone and Kazungula districts, where agriculture is a dominant and growing economic activity. In recent years, Livingstone has witnessed a growing focus of urban residents opting into farming as an additional income-earning opportunity, and rural residents in and around the periphery of Livingstone and Kazungula district areas predominantly focus on fishing, animal husbandry and crop farming as a livelihood strategy. With subsistence agriculture being a dominant economic activity for rural dwellers, it is worth exploring the opportunity for market linkages with the tourism industry. This aligns with Saarinen's (2007) argument that the significance of tourism is greater, particularly in marginalised peripheral or rural areas where the industry has been used as a catalyst for welfare, employment and economic growth. Employment and income are some of the inclusive growth attribute that sustainable market linkages, such as tourism-agriculture linkages, can create for small-holder farmers located in and around the Livingstone and Kazungula districts.

In this context, the aim of this study is to evaluate the interface between tourism development and agriculture in Livingstone, Zambia, with a focus on providing a better understanding of opportunities and barriers expected and experienced by small-holder farmers and how this contributes to inclusive tourism growth. This study then contributes to the emerging inclusive growth literature within the tourism domain using Livingstone, the tourist capital of Zambia, by employing qualitative and quantitative analysis focusing on the expectations and experiences of farmers participating in the tourism economy.

This paper draws from fieldwork that took place between May 2018 and October 2019 in the Livingstone and Kazungula districts of Southern Zambia. The study engages the qualitative approach to gain in-depth insight into the current tourism-agriculture market linkage. Primary data collection involves semi-structured questionnaires to 48 local farmers supplying the accommodation subsector in the two districts. Based on the demographic profile (Table 5.1), more than 50% of the respondents are women (58%), and the age range of participants is between 20 and 79, with most of the respondents (29, 2%) in the age group 50–59. While the majority (55%) of farmers had many years of farming experience (11–30) years, the study also included (35%) farmers with fewer (1–10) years of farming experience and only (10%) had over 30 years of farming experience. In terms of educational qualifications, most of the respondents (44%) have tertiary education and the majority (78%) of the farmers have been supplying the accommodation subsector for less than 10 years.

Table 5.1 Demographic profile of respondents (farmers)

Demographics characteristic	Category	Percentage
Age groups (years)	20–29	10.4
	30–39	18.8
	40–49	14.6
	50–59	29.2
	60–69	22.9
	70–79	4.2
Gender	Female	58.3
	Male	41.7
Highest level of qualification	Primary	27.9
	Secondary	27.9
	Tertiary	44.2
Nationality	Namibia	2.3
	Netherlands	2.3
	South Africa	2.3
	Zambia	93.0
Years in farming	1–5	17.5
	6–10	17.5
	11–20	27.5
	21–30	27.5
	31–40	10.0
Years as a supplier to the accommodation-subsector	1–9	78.6
	10–19	11.9
	20–29	9.5

All farmers in the study belong to a farmer cooperative. In some instances, through cooperative membership, individual farmers in the sample supply big hotels like Avani with 400 rooms. Farmers belonged to Mujala Women's Cooperative and Milimo Mibotu Women's Cooperative supply chickens, while Jackie Mwanampapa Cooperative members supply a variety of vegetables. In these cases, individual farmers put their produce together (a variety of vegetables and chickens) to supply Avani hotel through an intermediary who collects farm produce from a central location and delivers it to the hotel twice a week. This system creates a mutually beneficial outcome as opportunities are created for hotel demand to be shared across individual farmers within and across cooperatives and to ensure that quantities demanded by the hotel are met.

Information from farmers centred on socio-economic opportunities and challenges they expect and experience in the tourism food supply chain. The purpose of this information is to generate data that presents the challenges and opportunities posed to tourism-led inclusive growth, as is the focus of this study. Interviews were conducted with owners and/or employees of hotels, lodges and guest houses (including general managers, head chefs, procurement managers and food and beverage managers) and intermediaries in the food supply chain (known as lead farmers) in some cases.

5.3.2   Tourism-Agriculture Linkages

The accommodation sub-sector caters for different market segments, with international leisure travellers dominating the market and a growing domestic and regional business traveller segment. It is worth noting that the peak seasons in the destination are between March and September, which coincides with the high-water volumes of the Victoria falls and the summer holiday period for visitors from regions in the Northern hemisphere, largely leisure international travellers.

Food procurement practices among the sampled accommodation enterprises are informed by availability, quality and price. However, quantities sourced depend on occupancy at any given time. During peak season (around March to September), room occupancy across the sampled accommodation establishments averages 70–80%. There is a large concentration of local sourcing from multiple suppliers, as indicated in Table 5.2. In most cases, quantities supplied directly by the majority of small-scale farmers are minimal (including farmers supplying as a cooperative)

Table 5.2 Accommodation sub-sector and Agriculture linkage forms and characteristics

Linkage form	Characteristic
Intermediaries	Linked to individual local small-scale farmers, sources produce as far as Lusaka, 360KM away and imports mainly from South Africa. Serves as farm produce aggregator for accommodation sub-sector particularly for big hotels. Serves both small-holder farmers and hotels logistics involved in finding a market and in procurement, respectively.
Imports	A source of leakages. No stimulation of agriculture sector, incidence of food imports occurs among big hotels. Though minimal, large volumes are imported mainly from South Africa.
Supermarkets	Three leading foreign-owned chain supermarkets stock a mix of local and imported foods, including fruits and vegetables. Minimum stimulation of the agriculture sector as produce is imported and sourced centrally, and retail supermarkets demand bulk that individual local farmers do not currently meet. Products are more expensive and less fresh compared to local farmers and open markets
Farm cooperative	Stimulation of agriculture sector, minimal support from the government, no control of market or market information
Open local markets	Stimulation of agriculture sector, controlled by local municipal council and restricted time allocated for farmers to trade. Relatively high possibility for compromise of quality and health standards due to poor handling and storage of produce at open local markets.
Self-supply	Lowered food procurement costs
Informal with individual farmers	Fresh produce, good quality, sanitation and hygiene of acceptable standards, inability to meet quantities, not consistent with supply and largely seasonal suppliers

compared to what the farmers supply to the open local markets and supply and demand remains informal. However, indirectly, farmers supply the accommodation sub-sector through the open market as lodges, hotels and guest houses sampled in the study source produce from local open markets.

5.3.3 Inclusive Growth Outcomes Expected and Experienced by Farmers Supplying the Accommodation Sub-sector

A Likert scale of (1–5) was used to capture expectations and experiences of positive and negative outcomes in supplying the accommodation subsector. A Wilcoxon signed-rank test was then conducted between farmers' expectations of positive and negative outcomes and actual experiences of the same. Where the experiences of the farmers with regards to positive outcomes match their level of expectation, the farmers are satisfied and this reflects a positive contribution to inclusive tourism-driven growth. If the level of experience exceeds the level of expectation, the farmers are more than satisfied, reflecting a greater level of inclusivity. However, dissatisfaction occurs when the level of experience by farmers falls below their expectation level, denoting weakness in the agriculture-tourism linkage and exclusion to tourism-led growth. In relation to the challenges (negatives), if the level of experience exceeds the level of expectation, farmers are dissatisfied, and this indicates a limitation to the linkage and a threat to sustainable inclusive tourism-led growth. Perception scores of farmers in relation to positive and negative outcomes of local sourcing along with the respective p-values are depicted in Tables 5.3 and 5.4, respectively.

Perception scores of the positive inclusive growth outcomes pertaining to the level of expectation and the level of experience of farmers in Table 5.3 show

Table 5.3 Positive outcomes expected and experienced by smallholder producers

Positive socio-economic outcomes expected and experienced	W	p-value
Expected: Skills development – Experienced: Skills development	−0.902	0.367
Expected: Financial access – Experienced: Financial access	−0.944	0.345
Expected: Employment generation – Experienced: Employment generation	−1.034	0.301
Expected: Support on training to improve product quality – Experienced: Support on training to improve product quality	−1.221	0.222
Expected: Income generation – Experienced income generation	−2.610	0.009**
Expected: Improved access to markets – Experienced: Improved access to markets	−3.402	0.001***
Expected: Improved wellbeing – Experienced: Improved wellbeing	−3.243	0.001***
Expected: Linkage to networks – Experienced: Linkage to networks	−3.208	0.001***

Based on positive ranks. Perception scores are based on a 5-point rating scale. * = p < .05, ** = p < .01, *** = p < .001 (2-tailed test)

Table 5.4 Barriers expected and experienced by smallholder farmers

Challenges expected and experienced in the market linkage	W	p-value
Expected: Inadequate farming implements – Experienced: Inadequate farming implements	−4.309	0.000***
Expected: Inadequate transportation – Experienced: Inadequate transportation	−3.847	0.000***
Expected: Inadequate postharvest handling facilities – Experienced: Inadequate postharvest handling facilities	−3.492	0.000***
Expected: Poor growing conditions – Experienced: Poor growing conditions	−3.236	0.001***
Expected: Poor economies of scale – Experienced: Poor economies of scale	−3.055	0.002**
Expected: Farm labour deficit – Experienced: Farm labour deficit	−2.624	0.009**
Expected: Uncompetitive pricing – Experienced: Uncompetitive pricing	−2.256	0.024*
Expected: Seasonality – Experienced: Seasonality	−1.781	0.075
Expected: Lack of capital investment and credit – Experienced: Lack of capital investment and credit	−1.436	0.151
Expected: Inability to meet health standards – Experienced: Inability to meet health, sanitation and safety standards	−1.190	0.234
Expected: Inability to meet quality standards – Experienced: Inability to meet quality standards	−1.081	0.280
Expected: Lack of communication and exchange of information – Experienced: Lack of communication and exchange of information	−0.904	0.366
Expected: Late payments – Experienced: Late payments	−0.797	0.425
Expected: Language barrier – Experienced: Language barrier	−0.324	0.746
Expected: Inability to meet quantity demands – Experienced: Inability to meet quantity demands	−0.303	0.762
Expected: Marketing challenges – Experienced: Marketing challenges	−0.375	0.708

Based on positive ranks. Based on negative ranks. Perception scores are based on a 5-point rating scale. * = p < .05, ** = p < .01, *** = p < .001 (2-tailed test)

evidence that none of the positive indicators met the farmers' expectations. They rated their expectations higher than their experience, meaning their expectations were not met, as clearly evidenced from the calculated p-values. The null hypothesis states that the medians of expectations and experiences are equal against the alternative hypothesis that says the medians of expectations and experiences are different. Out of 8 positive inclusive growth outcomes tested in this study, four did not meet the farmers' expectations, notable significant differences exist between the level of expectation and the level of experience of farmers corresponding to income generation, improved access to markets, linkage to networks in the tourism sector and improved well-being have p-values that are less than 0.05. Therefore the null hypothesis is rejected and concluded that the median of the indicators mentioned are significantly different.

There is a notably high difference in improved access to markets, linkage to networks in the tourism sector and improved wellbeing where the experience was far less than what the farmers expected, especially in this regard. In respect these indicators, the findings seem to suggest that tourism-led growth does not contribute to these expected inclusive growth outcomes for small-holder farmers that participated in this study. However, while the median scores of farmers' expectations are higher than the median scores of experiences in relation to employment generation, skills development, access to loans /finances and support on training to improve product quality, statistically the mean differences are not significant. In this case, the null hypothesis is retained, and the conclusion that the means in these inclusive growth indicators are equal is made.

In addition to the positive outcomes tested in the study as indicated in Table 5.3 above, farmers reported positive experiences in the tourism-agriculture market linkage despite the low experience scores. This is the case particularly in relation to the ability to generate income and learning and upskilling opportunities that they have gained. Income generated from supplying the tourism market contributes to the livelihood of farmers' households. When asked what they used the revenue generated from supplying the tourism market directly, it was evident from their responses that all respondents (N = 48) showed that they use the income generated from supplying the tourism market on multiple aspects of their lives and livelihood, the most sited response (n = 28) was the use of income on household sustenance such as food, health, transport and other day to day needs. This is closely followed by paying school fees as well as reinvesting into the farming business. Respondents also highlighted that revenue generated from directly supplying the tourism industry was used to participate in community savings scheme or village banking initiative popularly known as SILC (Savings and Internal Lending Communities) Nsanzya (2022).

5.3.4 Challenges Expected and Experienced by Farmers Supplying the Accommodation Sub-sector

Table 5.4 above displays the perception scores of challenges which reveals that to a greater extent, farmers' levels of experience were higher than the median values of expectations. This implies that farmers' experience of the challenges was worse than what they had expected. However, there are some challenges where the farmers' experience was less than what they expected, as highlighted in the medians of the inability to meet health, sanitation and safety standards,; lack of communication and exchange of information, late payment by accommodation enterprises; Inability to meet quality standards; inability to meet quantity of demand, lack of capital investment and credit, marketing challenges, seasonality and language barriers. For these particular variables, the experience of farmers was better than they had anticipated.

Inadequate transportation, uncompetitive pricing, poor growing conditions, inadequate farming implements, poor economies of scale, and lack of or inadequate post-harvest handling facilities and farm labor deficit have p-values that are less than 0.05. Therefore, the null hypothesis (equal medians) is rejected, and the conclusion is that there is a significant difference between the medians of the expectation and experience of challenges mentioned. The p-value for seasonality is significant at a 10% level of significance; therefore, it can be concluded that there is a significant difference between what farmers expected and what they experienced medians.

It is evident that farmers do not experience the challenge of inability to meet health, sanitation, safety, and quality standards as the experience median scores are lower than the expected median scores (Table 5.4). This finding is confirmed by respondents from the accommodation sub-sector. However, it is interesting to note that, of the challenges that had no significant difference between the median scores of expectation and experience for farmers (marketing, lack of communication and exchange of information and the inability to meet quantity of demand), this demonstrates that farmers do not experience these challenges as limiting their participation and benefiting from the tourism sector. Yet, for the accommodation sub-sector, these were cited highly as some of the limiting factors to a sustainable market linkage between the two sectors, particularly that farmers are not able to meet the quantity of demand and they do not market their produce (Nsanzya, 2022). This reflects that while farmers experience these challenges as expected, the accommodation sub-sector amplifies the need to address these. One can conclude that there is a need for greater engagement and dialogue between the two sectors to identify challenges as they arise and understand the scale and impact of such barriers to a sustainable market linkage.

In addition to limitations reported in Table 5.4, farmers expressed other challenges relating to production or supply, marketing and lack of cooperation and partnerships among themselves as farmers, and with other stakeholders that are key to both the tourism and agriculture sectors. In relation to production factors, despite farmers having the potential to expand their farming enterprises, their efforts are constrained by limited finances to maintain farming business, expand farming activities and invest in farming implements. For most small-holder farmers, the cost of production (includes labour) is too high to remain sustainable, particularly in cases where labour from family members is not available. This is exacerbated by volatile prices of produce, exposing farmers to uncertain prospects of the market price in general and the accommodation sub-sector in particular. Lack of cooperation among farmers results in negative consequences affecting the sector. These include a lack of information sharing related to production, supply and market environment, which would benefit the agriculture sector and afford them opportunities to have a coordinated market approach to the accommodation sub-sector and other markets. This has far-reaching negative effects as farmers have no control of the market (product, price, demand). Saturation of the market, particularly with vegetables, frequently occurs as the majority of small-holder farmers tend to grow similar seasonal crops and compete for the market, thereby over saturating the market and resulting in undesirable effects for farmers in terms of the low product price and intensifying the risk of loss of income as well as food wastage.

The cost of managing animal and crop diseases is quite high for small-holder farmers with a minimal income base, which reduces profit margins. Other challenges that threaten sustained production include external factors such as wild animals, particularly for small-holder farmers living in the Katombora area where the Royal Chundu lodge is located. Climatic changes was cited as another major concern for farmers relying on rain-fed irrigation as this results in drought and water scarcity which negatively impacts on their production capacity. Furthermore, there are challenges experienced post-harvest, such as lack of transport for transporting produce from production sites to the market and facilities for packaging and storage such as cold rooms to ensure product quality is maintained and hygiene standards are observed.

Farmers cited major demand challenges, including limited quantities and low frequency of demand from the accommodation sub-sector resulting in small income margins from the tourism market segment, particularly when compared to demand from the general public via open local markets. This explains the low experience score on the inclusive growth outcomes. In addition, payment delays, uncompetitive pricing, misalignment of tourism peak season and farmers' production timelines were also cited as challenging experiences in the tourism supply chain. The lack of dialogue between the two sectors, unethical and unfair practices in the supply chain, particularly among intermediaries (payment delays to farmers and uncompetitive price), renders the tourism-agriculture market linkage undesirable for some.

Marketing related challenges experienced by farmers supplying the tourism market highlight lack of market information and limited market penetration into the accommodation sub-sector. In addition, farmers cited unfair competition and limited market for high-value crops (such as baby marrow, cauliflower, broccoli, lettuce and peas), which are largely consumed by the restaurant and hotel market and not popular among the ordinary local communities. In most instances, surplus high-value crops tend to go to waste as the hotel demand for most high-value crops is limited to guest occupancy at any given time. Some of the farmers felt that growing specific crops to meet hotel demand was risky, yielded low profits and was not economically viable given the time and effort required to produce such crops.

5.3.5   Strategies Proposed by Farmers to Address Unfulfilled Positive Experiences & Experienced Challenges

In order to strengthen the existing tourism-agriculture market linkage for inclusive growth, a number of strategies were proposed to address unfulfilled positive outcomes and significant challenges experienced by farmers (Table 5.5). Respondents provided justification for each of the strategies, and there is evidence that a single strategy can address more than one unfulfilled and significant outcome. For example, consistency in supply would also secure a sustained market with the tourism industry and give the farmer some level of security of market, so planning and coordination between the two sectors are important. In addressing the experienced challenges, there is a need for farmers to collaborate and partner within the agriculture

Table 5.5 Strategies to address challenges experienced by small-holder farmers

Unfulfilled inclusive growth outcomes & challenges	**Strategies proposed by farmers to address barriers to a sustainable and economically viable tourism market**
Skills development	Farmers to be educated on how to operate farming practice as a business; the current practice is that they search for the market at the point of near harvesting, and this is not sustainable in many instances resulting in produce going to waste. The government should identify farmers, train them, and help them obtain resources to run the farms properly, as farmers need capital to sustain and grow their farming enterprises. The state should supply guidance to monitor how farmers are farming to resolve challenges, and this can be done through the Ministry of Agriculture and the Department of Cooperatives, thereby strengthening the constitutional obligations of cooperatives.
Access to loans/ finance	Reduced interest rates for farmers to get bank loans.
Linkage to networks in the tourism sector	Establishment of food aggregators. Set-up can be done by farmers to supply the accommodation sector when they need food products and supply the chain stores when the accommodation sector demand is offseason. Retail chain stores to open up to local farmers
Improve well-being	Make the lead farmers stronger, and if they have the capacity to assemble all the products from the local farmers, it will contribute to the life of the local farmers.
Inadequate transportation	We need road infrastructure
Uncompetitive pricing	Fair pricing and prompt payment, farmer cooperatives coming together to regulate pricing by making the fixed price both with the sellers and buyers
Seasonality	Seasonality causes gaps in supply, and farmers should work on this challenge. While most quality is good, there is a need for more farms so that there are diverse crops that are widely available. This will address the challenge of limited supply. Government to identify farmers, train them and help them to obtain resources to run the farms properly. As farmers need capital to sustain and grow their farming enterprises, the recommendations made was to have reduced interest rates for farmers to get bank loans. Government is key, supporting farmers to be compliant and reduce ZRA taxes Agriculture needs to get support from their principals Supply guidance for people to monitor how farmers are farming to resolve challenges

(continued)

Table 5.5 (continued)

Poor growing conditions	The ministry of agriculture to provide extension services as basic support to farmers where we learn what different crops are favourable for specific soil types, crop management and disease and pest control. Currently we pay individually for extension services Ministry of Tourism and Arts, Bureau of Standards, Ministry of labour, Ministry of Agriculture to work together
Inability to meet quantity of demand	We need assistance with knowledge, we need assistance with financial support to meet hotel demand We would like to know more farmers as this gives us the advantage of choice, price and quality Farmers need to be aware of the needs of the accommodation sector Network of suppliers as a one stop place for hotels to buy from Farmers to have variety and be able to deliver Smallholder farmers to work with lead farmers to ensure consistency and quality for example Working together, assist with financial support, assist with knowledge so that we can expand and meet the hotel demand Consistency in supply would also guarantee market with the tourism industry and gives the farmer surety of market and price, so planning is important.
Inadequate farming implements	We need water pumps to reduce labour of watering by hand as we fetch water from the Zambezi river and it is not safe as there are crocodiles. If we have water pumps, we can expand our gardens. Also need pesticides as we only afford to buy a few pesticides at a time which do not cover the whole field We need transport and irrigation pipes to be mended Application of modern production facilities Power electricity, road infrastructure, bore holes for water security Funding opportunities to fence off farms as a measure to protect from wild animals
Poor economies of scale	Proper coordination, not equate small scale farmers to that of commercial farmers in terms of quantities demanded and formation of partnerships with other farmers
Marketing challenges	We need assistance to find a market as we are many farmers supplying the same lodge. At times, when you go there they tell you that they have already purchased from another farmer from within our community. Access to market information should be improved and Increase quantities given to small scale farmers Improve market linkages, information to farmers and good pricing Farmers and those from the tourism industry should have programmes that will enable them to interact Cooperation and partnerships for ease of monitoring of demand and supply Government to assist in promoting local procurement by the tourism sector as they did with Shoprite and Spur super markers Hotel owners should support small scale farmers adequately and get rid of all middlemen. There should be training often to communicate their expectations from farmers

(continued)

Table 5.5 (continued)

Lack of or inadequate post-harvest handling facilities	Need for funding opportunities to have packaging materials, transport and cold rooms
Lack of communication and exchange of information	Hold interactive meetings between small-scale farmers and the accommodation sub-sector. Cooperatives and the Ministry of Agriculture to assist in linking the two industries together. Improved coordination and communication both ways, farmers can meet demand if they know in advance.

sector and with the accommodation sub-sector and other players in the tourism economy.

Based on the results, support is needed for small-scale farmers in relation to supply or production-related challenges, including enhanced communication and education on operating farming practices as a business. The current practice is that they search for the market at the point of harvesting, and this is not sustainable or economically viable, which may result in an over-saturation of the market and inevitably a significant reduction in product prices and produce going to waste.

5.4 Conclusion and Recommendation

Drawing from the expectations and experiences of local small-holder farmers, the results of the study show that tourism-agriculture market linkage exists. However, based on the results, the linkages are weak and fragmented, resulting in low positive inclusive growth outcomes. This finding can be attributed to the extent of challenges related to supply, demand, marketing and the lack of support from government ministries responsible for tourism and agriculture. Tourism-driven growth can be interpreted to be inclusive as far as farmers participation in the tourism food supply chain is concerned, albeit with limitations as discussed above. When one considers the positive outcomes resulting from the tourism-agriculture linkage, these are low but could be improved as the potential for a strong and economically viable synergy is evident. This is indicated particularly by the resilience demonstrated by farmers – despite the many production challenges they experience.

While the government recognises the importance of linking tourism as an industry with other economic sectors as stated in the tourism policy (see Ministry of Tourism, 2015), one can conclude that there has been a minimal effort from the government through its ministries responsible for tourism and agriculture to enhance the existing tourism-agriculture linkage. However, there is potential for developing these linkages: e.g. the hotel industry demonstrated an inclination towards sourcing locally produced food and an appreciation of the quality and freshness of locally produced farm products. While reliance on food imports is minimal, local sourcing of food products is constrained by challenges and lack of knowledge of experiences and expectations of farmers and farmer production activities by the accommodation-sub-sector. In order to strengthen the market linkages, the government should take a leading role through policy-making and its ministries responsible for agriculture and tourism, farmer cooperatives and other stakeholders in the private sector. Farmer cooperatives and tourism associations in Livingstone could strengthen the synergy between tourism and agriculture sectors by addressing the challenges experienced. Consequently, this would create opportunities for inclusivity of tourism-led growth that would be more sustainable and socio-economically viable for small-holder farmers participating in the tourism-agriculture market linkages in the region.

References

Anderson, W. (2018). Linkages between tourism and agriculture for inclusive development in Tanzania: A value chain perspective. *Journal of Hospitality and Tourism Insights, 1*(2), 168–184.

Bakker, M., & Messerli, H. (2017). Inclusive growth versus pro-poor: Implications for tourism development. *Tourism and Hospitality Research, 17*(4), 384–391.

Berno, T. (2011). Sustainability on a plate: Linking agriculture and food in the Fiji Island tourism industry. In R. M. Torres & J. H. Momsen (Eds.), *Tourism and agriculture: New geographies of consumption, production and rural restructuring* (pp. 87–103). Routledge.

Biddulph, R., & Scheyvens, R. (2018). Introducing inclusive tourism. *Tourism Geographies, 20*(4), 583–588. https://doi.org/10.1080/14616688.2018.1486880

FAO (2012). Report on a scoping mission in Samoa and Tonga: Agriculture and tourism linkages in Pacific Island countries. FAO Sub-regional Office for the Pacific Islands.

Hall, C., & Lew, A. (2009). *Understanding and managing tourism impacts: An integrated approach.* Routledge.

Hall, C., & Page, S. (2014). *The geography of tourism and recreation* (4th ed.). Routledge.

Hampton, M., Jeyacheya, J., & Long, P. (2017). Can tourism promote inclusive growth? Supply chains, ownership and employment in Ha Long Bay, Vietnam. *The Journal of Development Studies, 54*(2), 359–376.

Hunt, H., Rogerson, C. M., Rogerson, J., & Kotze, N. (2012). Agriculture-tourism linkages in Botswana: Evidence from the Safari lodge accommodation sector. *Africa Insight, 42*(2), 1–17.

Jeyacheya, J., & Hampton, M. (2020). Wishful thinking or wise policy? Theorising tourism-led inclusive growth: Supply chains and host communities. *World Development, 131*, 104960.

Kavita, E., & Saarinen, J. (2016). Tourism and rural community development in Namibia: Policy issues review. *Fennia, 193*(3), 1–10.

Lacher, R. G., & Nepal, S. (2010). From leakages to linkages: Local-level strategies for capturing tourism revenue in Northern Thailand. *Tourism Geographies, 12*(1), 77–99.

Mbaiwa, J. (2005). Enclave tourism and its socio-economic impacts in the Okavango Delta, Botswana. *Tourism Management, 26*, 157–172.

McLachlan, S., & Binns, T. (2014). Tourism as a means for development in Livingstone, Zambia: Impacts among local stakeholders. *Australian Review of African Studies, 35*(2), 5–24.

Meyer, D. (2007). Pro-poor tourism: From leakages to linkages. A conceptual framework for creating linkages between the accommodation sector and 'poor' neighbouring communities. *Current Issues in Tourism, 10*(6), 558–583.

Meyer, D., Ashley, C., & Poultney, C. (2004). *Boosting local inputs into the supply chain.* Department for International Development.

Meyer, D. (2012). Pro-poor Tourism: From leakages to Linkages. A conceptual Framework for creating linkages between the accommodation sector and 'poor' neighbouring communities. *Current Issues in Tourism, 10*(6), 558–583.

Ministry of Tourism. (2015). *Tourism policy for Zambia*. Ministry of Tourism.

Ministry of Tourism and Arts. (2017). *2017 South-West Region tourist arrivals*. Government of Zambia.

Mitchell, J. (2010a). *An unconventional but essential marriage: Pro-poor tourism and the mainstream industry*. Overseas Development Institute.

Mitchell, J. (2010b). *Tourism in poor places- who gets what?* www.odi.org.uk/opinion/4717-tourism-poor-places-gets

Nsanzya, BMK. (2022). *Evaluating Tourism-Agriculture Linkages for Inclusive Growth in Zambia.*, PhD Thesis, University of Pretoria, South Africa.

Oyinlola, M. A., Adedeji, A. A., Bolarinwa, M. O., & Olabisi, N. (2020). Governance, domestic resource mobilisation, and inclusive growth in sub-Saharan Africa. *Economic Analysis and Policy, 65*, 68–88. https://doi.org/10.1016/j.eap.2019.11.006

Pillay, M., & Rogerson, C. (2013). Agriculture -Tourism linkages and pro-poor impacts: The accommodation sector of urban coastal KwaZulu-Natal, South Africa. *Applied Geography, 36*, 49–58.

Prideaux, B., Thompson, M., & Pabel, A. (2020). Lessons from COVID-19 can prepare global tourism for the economic transformation needed to combat climate change. *Tourism Geographies, 22*(3), 667–678. https://doi.org/10.1080/14616688.2020.1762117

Rogerson, C. (2013). Responsible tourism and local linkages for procurement: South African debates and evidence. *African Journal for Physical, Health Education, Recreation and Dance, 19*(Suppl 2), 336–355.

Rogerson, C., & Rogerson, J. (2020). Inclusive tourism and municipal assets: Evidence from Overstrand local municipality, South Africa. *Development Southern Africa, 31*(5), 840–854.

Rylance, A., Spencer, A., Mitchell, J., & Leturque, H. (2009). *Tourism-led poverty programme: Training module for agriculture*. International Trade Centre.

Saarinen, J. (2007). Tourism in peripheries: The role of tourism in regional development in Northern Finland. In B. Jansson & D. Muller (Eds.), *Tourism in high latitude peripheries: Space, place and environment* (pp. 41–52). CABI.

Saarinen, J. (2016). Political ecologies and economies of tourism development in Kaokoland, North-West Namibia. In M. Mostafanezhad, A. Carr, & R. Norum (Eds.), *Political ecology of tourism: Communities, power and the environment* (pp. 213–230). Routledge.

Saarinen, J. (2017). Enclavic tourism spaces: Territorialization and bordering in tourism destination development and planning. *Tourism Geographies, 19*(3), 425–437.

Saarinen, J. (2019). Communities and sustainable tourism development: Community impacts and local benefit creation tourism. In S. F. McCool & K. Bosak (Eds.), *A research agenda for sustainable tourism* (pp. 206–222). Edward Elgar Publishing.

Saarinen, J. (2020). Tourism and sustainable development goals: Research on sustainable tourism geographies. In J. Saarinen (Ed.), *Tourism and sustainable development goals: Research on sustainable tourism geographies* (pp. 1–10). Routledge.

Saarinen, J., & Rogerson, C. (2014). Tourism and the millennium development goals: Perspectives beyond 2015. *Tourism Geographies, 16*(1), 23–30.

Scheyvens, R., Carr, A., Movono, A., Hughes, E., Higgins-Desbiolles, F., & Mika, J. P. (2021). Indigenous tourism and the sustainable development goals. *Annals of Tourism Research, 90*, 103260. https://doi.org/10.1016/j.annals.2021.103260

Schilcher, D. (2007). Growth versus equity: The continuum of pro-poor tourism and neoliberal governance. *Current Issues in Tourism, 10*(2–3), 166–193. https://doi.org/10.2167/cit304.0

Spencer, J., Safari, E., & Dakora, E. (2014). An evaluation of the tourism value-chain as an alternative to socio-economic development in Rwanda, Africa. *African Journal for physical Health Education, Recreation and Dance, 20*(2), 569–583. http://erepository.mkiu.ac.rw/handle/123456789/6731

Torres, R., & Momsen, J. (2004). Challenges and potential for linking tourism and agriculture to achieve pro-poor tourism objectives. *Progress in Development Studies, 4*, 294–318.

Torres, R., & Momsen, J. (Eds.). (2011). *Tourism and agriculture: New geographies of production and rural restructuring*. Routledge.

Uduji, J. I., Okolo-Obasi, E. N., Onodugo, V. A., Nnabuko, J. O., & Adedibu, B. A. (2020). Corporate social responsibility and the role of rural women in strengthening agriculture-tourism linkages in Nigeria's oil producing communities. *Journal of Tourism and Cultural Change*. https://doi.org/10.1080/14766825.2020.1826500

UNWTO. (2020). *World tourism barometer*. https://www.e-unwto.org/toc/wtobarometereng/18/1

Van der Duim, R., Meyer, D., & Saarinen, J. (2011). Introduction: New alliances. In R. van der Duim, D. Meyer, J. Saarinen, & K. Zellmer (Eds.), *New alliances for tourism, conservation and development in Eastern and Southern Africa*. Eburon.

Welteji, D., & Zerihun, B. (2018). Tourism-agriculture nexus: Practices, challenges and opportunities in the case of bale Mountains National Park, Southern Ethiopia. *Agriculture and Food Security, 7*(8), 1–14.

World Bank. (2012). *Transformation through tourism: Development dynamics past, present and future*. World Bank.

Brenda M. K. Nsanzya is a Doctoral candidate at the Faculty of Economic and Management Sciences, University of Pretoria, South Africa. Her research interests are tourism linkages, tourism development impact, inclusive growth and community development.

Jarkko Saarinen is a Professor of Human Geography (Tourism Studies) at the University of Oulu, Finland, and Distinguished Visiting Professor (Sustainability Management) at the University of Johannesburg, South Africa, and Extraordinary Professor at the Tourism Management Division, Department of Marketing Management, University of Pretoria. His research interests include sustainable development, sustainable tourism, tourism-community relations and nature conservation studies.

Chapter 6
Insourcing the Indigenous Without Outsourcing the Story Teller: A Sustainable African Solution

Karen L. Harris and Christoffel R. Botha

6.1 Introduction

The heritage tourism sector has had to become a regulated domain given that it deals with cultural, historical and natural resources that must be protected and presented authentically, sensitively and sustainably. The purpose of the management and education of heritage tourist guides is twofold as they have a dual function as a mediator between the tourist (guest) and the indigenous community and their environment, as well as between the indigenous community (host) and the tourists (Weiler & Black, 2015a, b). In the light of this, this chapter argues that if the intention is to transform the tourism industry by making it more attractive and the sector more inclusive and sustainable, there needs to be less regulation and more innovation. It proposes to discuss the transformation of heritage tourism by including the indigenous story-teller (IST) within the ambit of the regulated tourist guiding sector.

This chapter focuses on a case study in the Northern Cape province of South Africa, home to one of the oldest indigenous peoples in the world, the //Khomani San (UP-DHHS, 2019). It will explain how a strategy can be developed to engage with, while at the same time monitoring the inclusion of ISTs within the tourism domain. Besides creating a procedure to incorporate these community voices into the tourist experience, the relationship is addressed between ISTs, tourist guides and tourists. It will argue that a mutually beneficial relationship needs to be in place for the effective involvement of ISTs in the broader heritage tourism realm. This will transform the tourist experience and understanding and enhance the position and place of the indigenous community within the heritage tourism sector in a sustainable manner.

K. L. Harris (✉) · C. R. Botha
Department Historical and Heritage Studies, University of Pretoria, Hatfield, South Africa
e-mail: karen.harris@up.ac.za; christoffel.botha@up.ac.za

© The Author(s), under exclusive license to Springer Nature
Switzerland AG 2022
J. Saarinen et al. (eds.), *Southern African Perspectives on Sustainable Tourism Management*, Geographies of Tourism and Global Change,
https://doi.org/10.1007/978-3-030-99435-8_6

We divide this chapter into four parts. Firstly, "tourism and the indigene" and how these two relate to one another; secondly, it considers "storytelling and tourism"; thirdly, how tourist guiding has been "regulated" and lastly, how the storyteller can be insourced.

6.2   Tourism and the Indigene

Pivotal to most tourist ventures is the desire to experience a "unique", "authentic", and "different" encounter. Since the end of the twentieth century, tourism has transformed from an essentially elitist and service-based industry to a more inclusive and pre-dominantly experience-based (Weiler & Black, 2015a, b). According to Alapuranen (2015), "experience-based logic" has dramatically transformed tourism. Experience-based logic is evident in tourism trends that have shifted from service-based to experienced-based, from mass consumerism to alternative niche-based experiences, from the conventional and general to the novel and authentic (Cooper, 2012). The drastic transformation to experience-based tourism left tourism enterprises needing renewal. Enterprises had to reformulate, sustainably develop and implement new novel and "authentic" ways to operate. The industry, therefore, embarked on schemes to create more meaningful tourist experiences and narratives for visiting tourists.

One key area which has emerged is the greater inclusion of indigenous communities within the collective tourism milieu. Indigenous peoples are culturally distinct societies and communities, generally located on the periphery of modern society, usually living in marginalised and isolated areas, destinations or regions. They often practice unique traditions to retain cultural, social, economic, environmental, and economic characteristics that keep them differentiated from the "dominant societies" in which they live, reside, and operate (Chronis, 2005). According to the United Nations Permanent Forum on Indigenous Peoples (2007) and the World Council of Indigenous People (1994), there are approximately 370 million indigenous peoples worldwide, spread across 90 countries (United Nations Permanent Forum on Indigenous Peoples, 2007; World Council of Indigenous People, 1994). Although only making up 5% of the global population, indigenous peoples not only account for about 15% of the worlds "extremely poor", they own, occupy and or may use almost a quarter (22%) of the world's surface area.

In many cases, indigenous peoples safeguard roughly 80% of the world's remaining natural and cultural biodiversity (United Nations Permanent Forum on Indigenous Peoples, 2007). International organisations have made many strides to recognise indigenous peoples globally in the last 50 years. However, vast theoretical and practical obstacles remain to successfully and sustainably incorporate indigenous communities and peoples into existing nation-states' economic markets, systems, and climates (Senehi, 2002). These obstacles are visible in practical barriers such as the limited access to previously owned land and tenure due to the effects of segregation, racism and the multivariate challenges to personal growth in these marginalised destinations (Fernandez-Llamazares & Cabeza, 2018a, b).

The addition of the indigene as part of the tourism package is not new. It has a long tradition of often forming part of tourism marketing strategies, particularly in former colonial destinations. However, while this was initially limited to the confines of museums, galleries, dioramas or staged cultural villages, it has now spread to a more authentic or real-time experience (Duminy, 2017). The indigene as part of the marketing strategy is apparent in a sampling of tourism in countries such as Canada, Australia and New Zealand in the global North and countries such as India, South Africa and China in the global South. In fact, in many cases, their tourism marketing highlights or hinges on these communities as a drawcard – particularly given their uniqueness to a particular destination. According to Butler and Hinch (2007), "indigenous cultures have become a powerful attraction for tourists" (Butler & Hinch, 2007). Sijer (2018) also reckons that indigenous tourism is a particularly fast-growing trend in the twenty-first century. Further evidence of this trend is the foundation of The World Indigenous Tourism Alliance (WINTA) some 8 years ago in 2012/13 (WINTA, 2019).

Including indigenous peoples within the tourism package has had mixed reactions and results. These range from potentially positive to adverse outcomes. On the positive side, there is economic upliftment, employment and poverty alleviation, and on the negative side impacts such as concerns over disappearing "uniqueness" and authenticity, commoditisation of culture and disruptions to the delicate socio-political structure generally inherent within these indigenous communities (Butler & Hinch, 2007). Both the advantages and disadvantages of including indigenous communities in the tourism domain need consideration to be a mutually beneficial and sustainable arrangement in holistic heritage management and education.

6.3 Storytelling and Tourism

Storytelling predates writing by millennia and is inherent in all communities globally at various levels (Chaitin, 2003; Rodil & Winschiers-Theophilus, 2007). Storytelling is a unique tool for passing on information about the history, heritage, and culture. Regardless of the origin of a particular story, storytelling remains a unique, powerful and dynamic interaction between the 'teller' and the 'listener' (Alapuranen, 2015).

Storytelling among indigenous communities has helped forge several fundamental principles within a specific context over time. Storytelling sustains communities, validates and expresses experiences, cosmologies, epistemologies, promotes the sharing of traditional knowledge, and nurtures ongoing relationships between community members, non-community members, the environment, and other applicable stakeholders (Iseke, 2013). However, indigenous people and their culture, heritage and traditions are under severe threat of destruction in today's modern world, particularly in areas surrounding traditional knowledge, authenticity, origin, cultural traditions, conservation and the rising fear of commoditisation and the demonstration effect within their culture (Haug, 2007).

Since its inception, storytelling has also been a "powerful feature" of the tourism industry. Successful tourist guides are "raconteurs who can weave history and contemporary facts together in a way that brings places and experiences to life" (Duminy, 2017). The ability of the tourist guide to inform in an entertaining fashion and bridge the divide between the visitor from the "outside" of the domain (the tourist) and the visited landscape (destination) remains pivotal to the binary tourist-tour guide experience. The story – narration, tale, chronicle or legend – shared by the tourist guide are the conduit between the tourist and the destination. Storytelling is thus integral to the very fibre of the business.

Storytelling has gained increasing attention as a global marketing trend and development tool in the tourism industry. Studies have shown that a "real" or "fictive" story associated with a destination or its peoples has the distinct ability to give the specific area an advantage over its competitors while still simultaneously providing the tourist with a more valuable and distinctive experience. Storytelling capitalises on the tourist's adherent need for a more authentic story behind the tourism product, along with the tourist's interest in how this "unique" story was acquired, created and then ultimately conveyed in an understandable and meaningful way (Chronis, 2012; Eskilsson & Hoghdahl, 2009). In this chapter, we suggest storytelling is also among the indigenes' many tangible and intangible attributes that can contribute to the experiential dimension of tourism. Storytelling persists as a foundational "mode for teaching and learning" for indigenous and other communities (Fernandez-Llamazares & Cabeza, 2018a, b).

Indigenous peoples are described as being "actually part of the site's fabric and thus interpret the value of the area within their cultural context" (Howard et al., 2001). They often tell stories about the place, what they have experienced, or what others told them happened at the site by adding a narrative element beyond what is observed but what is heard and imagined. However, in the process of incorporating the indigenous story, the indigenous storyteller or both, into the tourism offering, we have a "responsibility to protect the shared experiences of these […] indigenous communities, to value them and not to place them in a position of exploitation" (Duminy, 2017). Mossberg (2008) believes that tourist guides are crucial to facilitate this situation.

6.4   Regulated Tourist Guiding

The role of the tourist guide in the tourism industry is multi-faceted and multivariate (UP-DHHS, 2018). Tourist guides are perceived as representatives of their regions, cities, and countries qualified to guide tours. Many consider the tourist guide the "crucial link" between tourists and a country's multiple tourist attractions, nature, culture or adventure tourist sites (Van den Berg, 2016). This structured performance by a tourist guide directly impacts the success, sustainability, and nature of the tourist experience and hence the tourism sector. Thus, this extensive role of making a considerable contribution to how a tourism destination is perceived and promoting the sustainable use of all-natural and cultural resources by making visitors aware of an attraction's importance and vulnerability becomes the responsibility of the tourist guide (UP-DHHS, 2016).

The world over, tourism legislation formulates legislative and regulatory frameworks for sustainable development and management within a country. Implementing tourism legislation provides the protection and conservation of natural and cultural resources and the facilitation of the involvement of the private sector and local communities in tourism development activities. Tourism legislation reflects on the roles and responsibilities of all the essential stakeholders in the tourism industry by ensuring the rights of international and local tourists and the rights and obligations of participating businesses, inbound-outbound tour operators and all other concerned players in the tourism industry. Moreover, by implementing tourism legislation, a country can build consensus by consistency with a national tourism policy and long-term sustainability (Van den Berg, 2016).

Despite its apparent significance, legislation focussed explicitly on the tourist guiding sector has generally been neglected and inadequately regulated globally. Even where it does exist, tourist guiding legislation also has a surprisingly short history, only dating back to the mid-twentieth century. It is somewhat disconcerting that many countries still lack the appropriate policies and regulations to monitor their tourist guiding sectors, even though the international tourism sector is a massive contributor to annual global gross domestic profits (GDP) (WFTGA, 2011). Thus, front-line professionals (tourist guides) within the tourism industry are unregulated for the most part (UNWTO, 2017). Within the last two decades, developed and developing countries have started to realise that regulating this sector is vital to a country's tourism market in both the short and long term (Hall, 2008; UP-DHHS, 2015).

When considering the regulation of the tourist guiding domain, it becomes apparent how complex and encompassing it is. It involves several independent variables at both a direct and indirect level. Moreover, the tourist guide's integral and even indispensable role within the sector, no matter where tourism exists, makes the tourist guide a universal phenomenon. However, how they operate or are perceived, employed or even treated is not universal.

South Africa boasts a formalised tourism sector with a regulated tourist guiding division that predates many countries in the South African Development Community (SADC) region, on the African continent and internationally. Legislation about tourist guides dates back some three and a half decades and has, over time, witnessed numerous amendments and repeals. By law, registered tourist guides possess an accredited guiding qualification and a first-aid certificate. They agree in writing to adhere to a code of conduct and ethics. Furthermore, they train to have the necessary skills, insight and specialised knowledge to create a worthwhile experience for the tourist. Thus, in simple terms, registered South African tourist guides are professionals in their field (South Africa Information, 2013).

South Africa classifies tourist guides into three main types: cultural guides, nature guides, and adventure guides (Anon., 2017). In this context, a cultural tourist guide interprets the cultural heritage of sites to visitors by educating them about the different aspects of a particular area and is knowledgeable about that specific attraction, destination or location. (UP-DHHS, 2018) They may specialise in particular fields such as art, history, museums or historic buildings (National Department of Tourism, 2012). The requirements for guide registration, including what learning programme is needed, are the National Registrar of Tourist Guides (Anon., 2017). It is also important to note that no tourist guide may work without registering. An individual operating as a tourist guide but has not met the specific requirements concerning training, accreditation, and registration are officially in contravention of Tourism Act 3 of 2014 (National Department of Tourism, 2012). They will be liable for prosecution and a possible fine. Tour operators and other companies who employ illegal tourist guides can also be fined (Field Guides Association of South Africa, 2017).

While the regulation of the tourist guide may be necessary and beneficial, there may be another side to the coin. The categorisation of tourist guides in South Africa is not without precedent and appears to function well. However, there needs to be more flexibility in considering other possible avenues to enhance the sector and comply with the transformational requirement to make the industry more inclusive, diversified and accommodating of previously disadvantaged and marginal groups. Introducing the Indigenous Story Teller (IST) will have a range of positive returns. This local voice, with inherent knowledge, could enhance the authenticity of the tourist experience and contribute to the transformation and sustainability of the sector. It could also encompass the inclusion of indigenous knowledge in the broadest sense, adding to the uniqueness of the tourist's experience.

6.5 Insourcing the Storyteller – An African Sustainable Solution

The history of the African continent's people is steeped in oral tradition, also regarded as storytelling. The southern African Bushmen, who are essentially renowned for their rock art, are no exception to this. While their rock paintings are a national treasure dating as far back as 80,000–25,000 BCE, their storytelling abounds and persists today (Gilbert & Reynolds, 2011). Bushman stories or mythology have captured international attention from the first encounters with the outside world and continue to do so (Ross, 2009). Today they stand as one of the most over-researched peoples globally (Duminy, 2017).

Thus, in the Bushman case, not only do their stories remain integral to their indigenous settings, but they are also of relevance beyond that domain. The Bushmen are the quintessential storytellers. The final section will consider a case study of the //Khomani San in the Northern Cape province and the potential of storytelling within the context of the South African tourist guiding sector. Further, it will look beyond this community and region to consider the potential wider implications of establishing the IST within the broader South African tourism sector.

6.6 //Khomani Cultural Landscape, Northern Cape

The //Khomani San are one of the last surviving indigenous San communities in South Africa. Their living cultural landscape remains an essential aspect of South African culture post-1994 (Khomani San, 2018). Located in the northernmost corner of the Northern Cape province, the //Khomani Cultural Landscape is one of South Africa's newest declared World Heritage Sites (declared July 2017) and straddles the Kalahari desert landscape (Northern Cape Tourism Authority, 2018; UNESCO, 2018).

The//Khomani San's recorded history can be traced back to the eighteenth and nineteenth centuries, revealing peoples who overcame various events. These events include interactions with the southern migrating Bantu-speaking (black African) immigrants; European missionaries and colonial rule; forced removals in 1931; and land restitution in 1999 to the more recent establishment of the contemporary Heritage Park and the creation of the Bushmen Council in the mid-2000s (South African San Institute, 2009). This Council sees that all new developments within the area uphold the integrity and authenticity adherent with the //Khomani San's way of life and focuses on a critical set of ethics guidelines for each development project (South African San Institute, 2009). Today, the //Khomani San are actively involved in most designated World Heritage Site developments. They remain eager to explore even more potential projects to uplift the entire community and promote their unique way of life.

The IST can refer to an individual who belongs to an identifiable indigenous community with a shared origin, history, culture, and tradition. The IST can relate or share stories of their specific community's identity, heritage, history, culture and geographic context. The IST is particularly relevant to the tourism sector in that they add another dimension that provides a so-called authentic tourism experience. Instead of merely being told about the community, its origins, history and traditions, the IST makes it possible to engage with and experience first-hand the community member and participate in a "genuine or authentic encounter". Ideally, the IST can also present events or experiences without orchestrating scripting or choreography. Thus the //Khomani San, as ISTs, would share stories about their origins, history, traditions, and folktales as mentioned above.

However, this would not be possible within the South African tourist guide legislation. Strictly speaking, the regulations monitoring the tourist guide do not make it legal for such an encounter actually to take place. The IST does not have the prerequisite tourist guide accreditation or registration and therefore cannot "furnish […] information […] for reward" (National Department of Tourism, 2012). According to the regulations for tourist guiding, this would be in contravention of

Tourism Act 3 of 2014 (UP-DHHS, 2018). Tourist guide regulations exclude the IST from the tourism domain and, in a sense, flies in the face of transforming the sector.

The regulations contradict the very Act that it supports. Tourism Act 3 of 2014 is in place "to provide for the development and promotion of sustainable tourism for the benefit of the Republic [of South Africa], its residents and its visitors" with the preamble highlighting transformation (National Department of Tourism, 2012). Based on a lack of access to the tourist guide's formal training and accreditation process for multiple reasons, a //Khomani San IST would be excluded from participating in the tourism domain. Furthermore, even if the IST had access to some form of formal training, such as "soft skilling", avoiding conventional teaching methods, the authenticity of the storyteller and the indigenous storytelling experience would be undermined, making the whole encounter null and void. After all, Deacon and Dowson (1997) makes the point that generally, research shows that Africa is for tourists "a culturally pristine and authentic continent [...] the continent of local communities with authentic culture" (Deacon and Dowson, 1997). In their study of Bushmen, Hüncke and Koot (2012) likewise endorse this by stating that visitors expect authentic presentations of their everyday life and culture. Authenticity is therefore essential to a sustainable tourist experience.

For the //Khomani San to enter the sector as an IST, the tourism sector needs to become more accessible and encompassing. It needs to actively promote an inter-relationship between ISTs, the indigenous community, the tourist guide and the tourists themselves – realising the full potential of the authentic local voice within the overall tourist experience. However, this does not mean that the IST should be unmonitored or unregulated. Such a situation would only undermine the advantages of having a regulated tourist guiding sector which has proven to be the best way to ensure that the tourist receives the most efficient and worthwhile experience (UP-DHHS, 2015). There is too much at stake for that. Instead, specific roles and responsibilities already assigned to the culture/heritage guide should be enhanced and extended.

Assigning specific roles to the culture/heritage guide aligns with the point made by Weiler and Davis (1993). They contended that unlike 'adventure guides' and 'nature guides', 'heritage guides' need to balance tourism management, experience management and resource management. They emphasise the importance of the heritage guide (South Africa's equivalent culture guide) as going further and adopting a 'collective approach' to include the tourist, the environment, and community within the overall tourist narrative (Weiler & Davis, 1993). This collective approach sets a precedent for the South African tourist guide sector to consider augmenting the roles and responsibilities of the culture tourist guide as set out in the regulations.

Moreover, the recommendation is that the other attributes of the tourist guide, particularly the culture (and heritage) guide, need to be enhanced to position the IST within the industry without contaminating the original authentic story the IST presents. With this recommendation, the culture tourist guide only trains to act as a cultural mediator, intercultural communicator, facilitator, middleman, interpreter, intermediary, buffer, and a co-creator of memorable experiences for tourists engaging with ISTs. They essentially create the platform for ISTs at a particular tourist attraction to 'tell' their authentic 'story' to the tourists. Doing this aligns with the various calls of the past 30 years by scholars to include indigenous voices in the tourism industry and directly address the lack of inclusivity within the South African tourism market. Promoting ISTs as an authentic 'tourism product' will fill an evident need within the South African market and simultaneously create a diversified strategy for the IST to form an integral part of the culture tourist guide domain and the tourist guiding sector.

The IST is facilitated by positioning the accredited and registered cultural guide. Guiding qualifications, in South Africa, for a culture guide are made up of a collection of unit standards. In this context, unit standards describe" sets of principles and guidelines by which records of achievements are registered to enable national recognition of acquired skills and knowledge, thereby ensuring an integrated system that encourages life-long learning". (SAQA, 2017). The recommendation is that an additional (possibly optional) unit standard be added to the culture tourist guide. The purpose of the unit standard will enable the culture guide to facilitate and provide a platform for the inclusion of the IST in the tourism sector. The culture guides are elevated to a position where they can promote the inclusion of the IST in the tourism offering. Through the unit standard, they will be made aware of the sensitive and possible sacred nature of the encounter with the IST. Culture guides will also need to play the role of a conservator in ensuring that the IST and their landscape is preserved and sustained. They must be able to mediate between the IST and the tourist group as an intercultural communicator and interpret linguistic and other indigenous norms. The cultural guide must also comply and execute the payment of the IST according to predetermined fees. In essence, the cultural tourist guide becomes the co-creator of the sustainable indigenous storytelling experience in this role.

However, in extending the cultural tourist guide's role to accommodate ISTs, it is vital to note that the IST is not exclusively the//Khomani San domain. The concept adheres to a diverse collection of local voices already situated throughout the Northern Cape province that can directly contribute to the authenticity and integrity of a particular tourist attraction, destination and site. This model can also be applied generically throughout South Africa with the range of indigenous and marginalised voices that abound.

As a labour-intensive industry and one of the potential critical drivers of the South African national economy, tourism should contribute to job creation, poverty alleviation and black economic empowerment (Anon., 2018a, b). Yet to deliver on this mandate, the regulated tourism and tourist guiding sectors need to address those issues that obstruct its expansion and job creation. In a nutshell, tourist guiding regulations need to be more flexible and creative to address the challenges and accommodate the vibrant and changing nature of the tourism domain. Flexible and creative tourist guiding regulations are in line with the statement made by the South African State President when he stated that for the future development of the tourism industry: we should "take further measures to reduce regulatory barriers" (Anon., 2018a, b). The requirements of the culture guide need to be adapted and changed in the regulations to give them the jurisdiction to accommodate the IST within the tourism offering. There should be a constant reappraisal of the sector and the need for renewal to be sustainable.

References

Alapuranen, P. (2015). *Storytelling in experience creation: Case Kaisus Lappland. MA dissertation*. Lapland University of Applied Sciences.

Anon. (2017). Tourist guides in South Africa. *Tourism Tattler*, 1–2.

Anon. (2018a). New president urges South Africans to focus on tourism in key speech. *Tourism Update, 1*.

Anon. (2018b). South Africa can be tourism hub of the world – Ramaphosa. *Traveller, 24*, 1.

Butler, R., & Hinch, T. (2007). *Tourism and indigenous peoples: Issues and implications*. Routledge.

Chaitin, J. (2003). Stories, narratives and storytelling. Retrieved from http://www.beyondintractability.org/narratives. August 2020..

Chronis, A. (2005). Constructing heritage at the Gettysburg Storyscape. *Annals of Tourism Research, 32*(2), 389–406.

Chronis, A. (2012). Tourists as story-builders: Narrative construction at a heritage museum. *Journal of Travel and Tourism Marketing, 29*(1), 444–445.

Cooper, C. (2012). *Essentials of tourism*. Pearson Limited.

Deacon, J., & Dowson, T. A. (1997). *Voices from the past: Xam Bushmen and the Bleek and Lloyd Collection (The Khoisan Heritage Series)*. Witwatersrand University Press.

Duminy, E. (2017). Protecting authentic storytelling in tourism. *Tourism Opinion South Africa, 1*.

Eskilsson, L., & Hoghdahl, E. (2009). Cultural heritage across borders? – Framing and challenging the Snapphane Story in southern Sweden. *Scandinavian Journal of Hospitality and Tourism, 9*(1), 65–80.

Fernandez-Llamazares, A., & Cabeza, M. (2018a). Recovering the potential of indigenous storytelling for conservation practice. *Conservation Letters, 11*(3), 1–4.

Fernandez-Llamazares, A., & Cabeza, M. (2018b). Rediscovering the potential of indigenous storytelling for conservation practice. *Journal of the Society of Conservation Biology, 11*(3), 1–12.

Field Guides Association of South Africa. (2017). Retrieved from: http://www.fgsa.co.za. September 2020.

Gilbert, E., & Reynolds, J. T. (2011). *Africa in world history*. Pearson.

Hall, C. M. (2008). *Tourism planning: Policies, processes and relationships*. Prentice-Hall.

Haug, M. (2007). *Indigenous people, tourism and development? The San peoples involvement in community-based tourism. MA dissertation*. University of Tromso.

Howard, J., Thwaites, R., & Smith, B. (2001). Investigating the role of the indigenous tour guide. *Journal of Tourism Studies, 12*(2), 32–33.

Huncke, A., & Koot, S. (2012). The presentation of Bushmen in cultural tourism: Tourists' image of Bushmen and the tourism provider's presentation of (Hai//om) Bushmen at Treesleeper Camp, Namibia. *Critical Arts, 26*(5), 671–689.

Iseke, J. (2013). Indigenous storytelling as research. *International Review of Qualitative Research, 6*(4), 559–577.

Khomani San. (2018). Retrieved from: http://www.khomanisan.com. September 2020.

Mossberg, L. (2008). Extraordinary experiences through storytelling. *Scandinavian Journal of Hospitality and Tourism, 8*(3), 195–210.

National Department of Tourism. (2012). Retrieved from: http://www.tourism.gov.za. November 2020.

Northern Cape Tourism Authority. (2018). Northern Cape: #Khomani Cultural Landscape, World Heritage Site. Retrieved from: http://www.experiencenortherncape.com. November 2020.

Rodil, K., & Winschiers-Theophilus, H. (2007). *Indigenous storytelling in Namibia. Conference paper – International Conference on Culture and Computing.* Kyoto University.

Ross, R. (2009). *A concise history of South Africa.* Cambridge University Press.

SAQA South African Qualifications Authority. (2017). Retrieved from: http://www.saqa.org.za. July 2020.

Senehi, J. (2002). Constructive storytelling: A peace process. *Peach and Conflict Studies, 9*(2), 41–57.

Sjier, G. (2018). To see or not to see – The impact of indigenous tourism. *Transforming Tourism News*, 1-4.

South Africa Information. (2013). Retrieved from: http://www.southafrica.info. September 2020.

South African San Institute. (2009). Retrieved from: http://www.san.org.za. December 2020.

UNESCO (United Nations Educational, Scientific and Cultural Organisation) (2018). Retrieved from: http://www.unesco.org. December 2020.

United Nations Permanent Forum on Indigenous Peoples. (2007). Retrieved from: http://www.un.org/development/desa/indigenouspeoples/. September 2020.

UNWTO (World Tourism Organisation). (2017). Retrieved from: http://www.unwto.org. October 2020.

UP-DHHS University of Pretoria – Department of Historical and Heritage Studies. (2015). *Harmonisation of tourist guide training regulations and standards in southern Africa (Phase III).* University of Pretoria.

UP-DHHS University of Pretoria – Department of Historical and Heritage Studies. (2016). *Harmonisation of tourist guide training in Southern Africa.* University of Pretoria.

UP-DHHS University of Pretoria – Department of Historical and Heritage Studies. (2018). *A policy review of the tourist guiding sector in South Africa.* University of Pretoria.

UP-DHHS University of Pretoria – Department of Historical and Heritage Studies. (2019). *The Indigenous Story Teller (IST): The Northern Cape as a case study.* University of Pretoria.

Van den Berg, L. M. (2016). *Tourist guiding legislation: South Africa, Australia and Canada in a comparative perspective. MA dissertation.* University of Pretoria.

Weiler, B., & Black, R. (2015a). The changing face of the tour guide: One-way communicator to choreographer to co-creator of the tourist experience. *Tourism Recreation Research, 40*(3), 1–2.

Weiler, B., & Black, R. (2015b). *Tour guiding research: Insights, issues and implications.* Channel View Publications.

Weiler, B., & Davis, D. (1993). An exploratory investigation into the roles of the nature-based tour leader. *Tourism Management, 14*(2), 91–98.

WINTA. The World Indigenous Tourism Alliance. (2019). Empowering opportunity through tourism. Retrieved from: http://www.winta.org. August 2020.

World Council of Indigenous People. (1994). Retrieved from: http://www.cwis.org. September 2020.

World Federation of Tourist Guide Associations. (2011). Retrieved from: http://www.wftga.org. December 2020.

Karen L. Harris is a Full Professor and Head of the Department of Historical and Heritage Studies at the University of Pretoria as well as the Director of the University Archives. She holds a PhD in History and is an accredited tourist guide. Her research interests include innovative methodologies for the teaching of History and Heritage and Cultural Tourism.

Christoffel R. Botha is a Lecturer in Heritage and Cultural Tourism in the Department of Historical and Heritage Studies at the University of Pretoria, South Africa. He is also a researcher for the National Department of Tourism and an accredited tourist guide. His research areas include tourist guiding and tourism policy in the global South.

Chapter 7
Assessment of Costs and Benefits of Joint Venture Partnerships in Community-Based Tourism Between the Private Sector and Goo-Moremi Residents, Botswana

Bontle Elijah, Naomi N. Moswete, and Masego A. Mpotokwane

7.1 Introduction

Tourism is the world's largest industry employing millions of people (Hannam & Knox, 2010; WTTC, 2017). The industry has opened opportunities for the creation of employment, environmental education projects, preservation, infrastructure development, thereby improving the lives of rural dwellers, especially in the developing world (UNWTO, 2018; Van Vuuren, 2019; WTTC, 2019). Many regions have resorted to tourism in their economic development because of the associated benefits on people and their environments. For the most part, research on socio-cultural positive impacts of tourism includes knowledge and skills transfers (Moswete et al., 2009), poverty alleviation (Snyman, 2012a; Spenceley & Goodwin, 2007), empowering communities (Moswete & Lacey, 2014; Okech, 2007), renewal of cultural traits such as music, dance, beadwork (Lowenthal, 2006; Saarinen, 2010).

Community-based tourism (CBT) refers to initiatives that emphasise the development of local communities (Blackstock, 2005) and allow for residents to have substantial control over and involvement with its development and management, and a major proportion of the benefits remaining within communities (Moswete & Thapa, 2018; Ndubano, 2000; Salazar, 2012). Notably, CBT promotes community participation (Blackstock, 2005) and aims at generating benefits by allowing tourists to visit, interact with, and learn about, host populations culture and natural heritage (Lowenthal, 2006; Moswete & Lacey, 2014; Saarinen, 2010; Salazar, 2012; Sebele, 2010). Although CBT is associated with economic, socio-cultural, and environmental

B. Elijah · N. N. Moswete (✉) · M. A. Mpotokwane
Department of Environmental Science, University of Botswana, Gaborone, Botswana
e-mail: moatshen@ub.ac.bw

© The Author(s), under exclusive license to Springer Nature
Switzerland AG 2022
J. Saarinen et al. (eds.), *Southern African Perspectives on Sustainable Tourism Management*, Geographies of Tourism and Global Change,
https://doi.org/10.1007/978-3-030-99435-8_7

benefits, there is potential that there might be short – and long-term negative impacts on the host community (Moswete & Lacey, 2014; Snyman, 2012b; Stone, 2015). In some cases, CBT has been developed at the expense of the wellbeing and integrity of the destination community because monetary benefits are more highly regarded.

In Botswana, the CBT initiative was introduced following the realisation that enclave tourism, which provides exclusive facilities for the tourists that are far removed from the quality of services enjoyed by the local economy, does not benefit remote rural communities (Mbaiwa, 2005; Mbaiwa & Hambira, 2020). The Government of Botswana defines CBT as "tourism initiatives that are owned by one or more communities or run as joint venture partnerships with the private sector with equitable community participation, as a means of using natural resources in a sustainable manner to improve their standard of living in an economic and viable way" (Government of Botswana (GoB), 2007, p. 5). It is through the CBT approach that local people are encouraged to form a joint venture partnership with a private tourism operator and jointly run/operate a community-based organisation (CBO), also referred as Trusts from which they can collectively benefit (see GoB, 2007; Rozemeijer, 2001). This paper utilises a social exchange theory with an empowerment approach to understand the effects of joint venture partnership in community-based tourism in Botswana. The objective of this study was to assess the costs and benefits of joint venture partnerships between the private tourism operator and the residents of Goo- Moremi village in community tourism. The main aim of the research was to establish an understanding of the joint venture partnership between the community CBO/Trust and the private tourism operator at Goo Moremi-village. The research questions for the study are:

  (i) Is the joint venture partnership (JVP) between the Goo-Moremi residents and the private operator a benefit or disadvantage?
 (ii) What community tourism benefits have the residents derived from the JVP?
(iii) What are the costs/challenges the community faces?
 (iv) Does the joint venture partnership 'empower' the community?
  (v) Who between the two in the relationship has more influence and power?

The residents' views were analysed based on the following four research questions (variables): knowledge about the community CBO/Trust, level of commitment, tourism benefits and costs (challenges).

7.2   Community-Based Tourism in Botswana

7.2.1   Institutional Background: Community Based Natural Resource Management (CBNRM)

Tourism policies and strategies in Botswana (BTDP, 2003; GoB, 1990, 2007) put a strong emphasis on the need for rural communities to own tourism-related businesses and derive benefits from them (Mmopelwa & Mackenzie, 2020; Moswete &

Thapa, 2016; Stone, 2015). The government is in favour of promoting a tourism sector that encourages citizen participation in community-based tourism (GoB, 2007). The government adopted community-based natural resource management (CBNRM) principle to encourage local people to be involved in the tourism industry. The CBNRM is an approach born out of the Zimbabwe's Communal Areas Management Programmes for Indigenous Resources (CAMPFIRE) (Child et al., 1997). Hence, the CBNRM concept was framed as an initiative or strategy for nature conservation and economic development (Cassidy, 2000; Child et al., 1997). It was an approach introduced in the late 1980s when the United States Agency for International Development (USAID) and the government of Botswana embarked on a joint Natural Resource Management project (NRM). A joint venture, in this case, was defined as a business activity undertaken between two or more partners for their mutual benefits (Cassidy, 2000; DWNP, 1999; Rozemeijer, 2001; Snyman, 2014). The local people are partners in a community joint venture, with user rights to existing natural and cultural resources in the area, with an established private operator that recognises an area's potential for tourism (Snyman, 2012b). The idea of a joint venture partnership between community and private tourism operator was built to yield positive results on the conservation of natural and cultural resources (Cassidy, 2000; Mbaiwa, 2011, 2013; Rozemeijer, 2001; Snyman, 2014). At most, the CBNRM approach rests on the recognition that local people must be actively involved in the management and use of resources (Cave & Negussie, 2017; GoB, 2007; Howard, 2003; Keitumetse, 2016; Mbaiwa, 2011; Saarinen et al., 2014; Snyman, 2014; Tomaselli, 2012).

7.2.2 Community-Based Cultural Tourism

Contextually, CBNRM based cultural-heritage tourism in Botswana is associated with ecotourism since it is built on similar principles. In essence, ecotourism is regarded as a Siamese twin between nature and cultural resources (Weaver, 2001). As well, conservation and ecotourism are deemed as effective partners in all types of protected areas (Eagles et al., 2002). Operationally, ecotourism is a type of tourism that is practised in sensitive and pristine areas where ecosystems are protected, which intend to have a low impact on natural and cultural environments (Cave & Negussie, 2017; Honey, 1999; Stevens & Jansen, 2002). It is often small scale and intended as an alternative to commercial 'mass' tourism (Weaver, 2001).

The benefits of cultural heritage are numerous (Cave & Negussie, 2017; Manwa et al., 2016) and include, but are not limited to, conservation of endangered species (flora and fauna), preservation of culture and traditions (Dichaba, 2009; Mbaiwa & Sakuze, 2009; Saarinen, 2012), preservation of heritage sites and monuments (Keitumetse & Nthoi, 2009), protection of wetlands of universal importance such as the Okavango Delta (Mbaiwa & Hambira, 2020). Thus, Botswana's CBNRM cultural heritage tourism is premised on the principles of sustainable tourism. Sustainable cultural heritage tourism, therefore, is a form of tourism which is guided by safeguarding and preservation of tourism resources (GoB, 2007; Keitumetse, 2016; Lowenthal, 2006), while observing the wellbeing of local people (Honey, 1999; Saarinen & Rogerson, 2015). A CBNRM Joint Venture Partnership (JVP) requires a community and a private company to work together, sharing the risks and responsibilities of a co-managed enterprise. The partnership generally offers the community more decision-making power and training (DWNP, 1999; GoB, 2007; Van der Jagt & Gujadhur, 2002). In rural Botswana, the JVP success stories include Khwai Development Trust; Sankuyo Tshwaragano Management Trust and Okavango Community Trust (Mbaiwa, 2005: 45–47; 2013 and Snyman, 2014: 123). However, not all JVP CBO/Trust are successful – hence this study sets to enquire about the costs and benefits of a joint venture partnership in community cultural tourism in the case of Goo-Moremi community in the Tswapong region of the Central District in Botswana.

7.3 Moremi Community-Based Organisation and Tourism

The residents of Goo-Moremi village formed a community-based organisation (CBO/Trust) known as Goo-Moremi Mannonnye Conservation Development Trust (GMMCT). The CBO/Trust was formally registered in July 1999 (White, 2001) with the help of a Botswana Non-Governmental Organisation (NGO) – the Kalahari Conservation Society (KCS). The area of operation was Goo-Moremi Mannonnye Conservation Area – a protected region encompassing the Goo-Moremi Gorge (GoB, 2001). The GMMCT was created to protect and preserve cultural heritage and natural resources of Goo-Moremi. When the area was gazetted for conservation, the residents of Goo- Moremi were introduced to community-based cultural tourism. Several organisations, including the Department of Tourism executive eco-tourism committee, the National Museum and the Kalahari Conservation Society, visited the community Goo Moremi on several occasions to sensitise them about the need to conserve, sometimes preserve, but at all times initiate conservation-based tourism in their area. The residents of Goo-Moremi were encouraged to venture into cultural-heritage tourism in their localities and were made aware of the community and individual benefits such as employment creation opportunities, income generation and a sense of ownership and pride in their own cultural traditions and heritage resources (Dichaba, 2009; White, 2000).

7.4 Methodology

7.4.1 Study Area: Context

Goo-Moremi village is located about 45 km east of Palapye on the northern edge of the Tswapong Hills. The village is situated at the foot of the greater Tswapong hills, at the mouth of the conservation zone known as Goo-Moremi Manonnye

Conservation Area (GMMCA) (GoB, 2001). The population of the Goo-Moremi village was 597 in 2011 (Statistics Botswana, 2016), with a high proportion of elderly people of 60 years and older. Many young people have migrated to major villages, towns and cities elsewhere in the country for greener pastures (Statistics Botswana, 2016; *conversation with the village leadership, 2015*). Agriculture is the dominant source of livelihood, with 50% of the people who practice crop and animal farming (GoB, 2001; Statistics Botswana, 2016). There is minimal tourism activity taking place within the village based on the cultural heritage resource of the Moremi Gorge (GoB, 2001; Mbaiwa, 2011; Siphambwe et al., 2017), and the community regards the gorge highly due to its rich history and their ancestral spirit associated with it (Dichaba, 2009). According to Mbaiwa and Siphambe, (Mbaiwa & Siphambe, 2019), Goo-Moremi residents have been culturally resilient and did not bend to the wishes of the tourism industry and holidaymakers at the expense of polluting their cultural values.

The village is endowed with natural and cultural heritage resources which include the Goo Moremi Gorge. The gorge is located about 14 km from the Palapye-Bobonong turn off and is approximately 4 km south of Goo-Moremi village *Kgotla*. The Gorge and its surrounding cultural and natural landscapes were declared a national monument in 2005 (Geoflux, 2009), from which date it was protected and managed by the government of Botswana through the Department of National Museums, Monument and Art Gallery (GoB, 2001).

With reference to local history, the Goo-Moremi Gorge is regarded as a home, or resting place, of the spirits of the Goo Moremi forefathers (ancestors) (Dichaba, 2009; Siphambwe et al., 2017; White, 2001). Culturally, the ancestral spirits are in control of the lives of the people of Goo-Moremi village through a prescription of taboos and cultural beliefs (Dichaba, 2009; GoB, 2001; Mbaiwa & Siphambe, 2019). The belief system forms the sacred culture and dignity of the people of Goo-Moremi village (GoB, 2001; Mbaiwa, 2011). There are well known and respected taboos and beliefs associated with 'the Gorge' (Geoflux, 2009; personal communication with village elders, 2016). The village residents believe that the ancestors can bring punishment in the form of bad luck to those who disregard the taboos and beliefs (personal communication with the village elder).

7.4.2 The 'GORGE' as an Ecotour and a Place for Tourism Activities

A gorge is defined as a "rocky-walled, steep-sided deep narrow valley" (Clark, 1998, p. 173), and has a V-shaped cross-section. A gorge is formed because of a change in rock type at the base of a waterfall as the pressure and erosive action of the flowing water causes the softer rock beneath it to erode, thereby forming a depression or plunge pool (Adams & Wyckoff, 1971; Bayliss, 1995, p. 81). Thus, the Moremi Gorge is typical as it is characterised by a vertical-walled valley which could have been cut by a stream following master joint and the subsequent edges of more resistant sandstone layers, making it like ribs. According to White (2001), the Moremi Gorge is one of the few places and geo-sites in Botswana with perennial water streams that pours into different pools.

Locally, Goo-Moremi Gorge represents a unique tourism destination in Central Botswana (GoB, 2001). Recently, there has been an increase in international and domestic tourists to the area, some of whom stay overnight in permanent thatch-roofed and tented chalets. Other groups of visitors choose wilderness non-motorised camping where they pitch their own or hire tents (personal communication, with the Manager of the CBO/Trust). According to early researchers in the area (Geoflux, 2009) camping was not very attractive in the early 2000s (White, 2001), but increased over the years due to the provision of ablutions which include flush toilets and hot shower facilities and braai/barbecue place at the camping sites (see Rossman & Rallis, 2017).

For tourism, visitor activities offered at the destination include but are not limited to hiking, walking, bird watching, game viewing, photography. There were no organised game viewing tours, but guided trail tours were offered at the time of the study (personal discussion with the guide and staff). The gorge is renowned for its waterfalls, perennial streams, and pools (natural ponds). There are about 5–6 fasci-nating pools that constitute the most part of the Moremi Gorge, and the popular one which holds water all year round is pool-gorge number three.

The Goo-Moremi Manonnye Conservation Area attracts different segments of the leisure and tourism markets both domestic and regionally, which include educa-tional tours, religious groups, cultural tourists, wedding photoshoot groups, camp-ers (Geoflux, 2009; Mbaiwa, 2011; Njadingwe, 2015). The monument is patronised by day-trippers, excursionists, climbers, adventure tourists and at most, the gorge is a drawcard to hundreds of domestic tourists, and the international tourists come mainly from South Africa.

Archaeological research conducted in the area discovered a wealth of history in Goo-Moremi, especially at the Moremi Gorge (Dichaba, 2009; Geoflux, 2009; Njadingwe, 2015). Historical and archaeological remains found at Goo-Moremi Gorge include but are not limited to the graves of Kgosi Mapulane, the chief who led the people of Goo-Moremi Village from South Africa to Botswana, and his sons. The archaeological remains (stone foundation) of the old village, school and area are dotted with burial mounds or graves of past village leadership, which form a significant heritage trail at the Goo Moremi (site visit in 2017). According to Mbaiwa (2011), the strategic location of Moremi village, the wealth of cultural heri-tage resources, history, the 'Gorge' and in particular the Goo Moremi village, have become one of the main tourism destinations and a unique attraction in Tswapong area and the Central District region of Botswana. Day trippers, overnight stay tour-ists and other tourism segments such as church groups, wedding groups, and school tours have increasingly frequented the gorge (personal communication with staff, 2015).

7.4.3   Data Collection and Analysis

Data were collected from July – August of 2013 and March 2014 among residents of Goo-Moremi village. Additional information was collected in 2016. During preparation for data collection, protocols for a social survey (Bernard, 2000; Creswell & Clark, 2011; Rossman & Rallis, 2017) were adhered to before the work commenced at the study site. Introductory meetings were held with the village chief and other key members of the village, such as the Village Development Committee chairperson, CBO/Trust Chair, and Goo- Moremi resort manager, to explain the objectives of the study (Groves et al., 2004; Lavraks, 2008) and thereby seek their consent. A semi-structured questionnaire with open and close-ended questions was used to collect data for the study. Fifty per cent household heads or representatives were selected through convenience sampling technique (see Bernard, 2000), while seven key informants (n = 9) were selected through snowball and purposive sampling for face – to face – interviews (Flick, 2011; Groves et al., 2004; Lavraks, 2008; Veal, 2006). An interview guide was used to solicit information from, among others, the village headmen, some representatives of the CBO/Trust, Village Development Committee (VDC), local guides and the site manager. Participatory observation and informal discussions about tourism and the gorge took place during trail/walking tours. Extra information for this study was also obtained through participatory observation and facilitation at the heritage marathon workshop by the second researcher held at Goo Moremi in 2017. Secondary information was sourced from the Goo Moremi Manonnye Conservation Area consultancy report by White (2001).

Based on descriptive data analysis, the majority (84%) of the respondents from the household survey were females. In rural Botswana, women tend to remain home to tender for household chores and take care of children while men work in the farms or in other major towns and villages. The modal age was in the 36–40 age groups (34%), followed by 26–30 age groups (28%) and 41 and above (18%), while the least represented were in the age group of 31–35 (4%) (Table 7.1).

Fifty-eight of the respondents had secondary education, whilst 12% had no formal schooling. Eight per cent had attained tertiary/university education. More females (70%) had formal schooling when compared to males (14%). Thus, the illiteracy level appears to be slightly high at (12%), representing six persons. The majority, 82%, of the respondents are literate, considering that they have gone through formal schooling in their lifetime. The nine key informants age ranged from 21 to 72 years. There were five males and 4 females. All the key informants had formal schooling. About three of them were not indigenous Moremi village residents but were there for employment purposes.

For data analysis, a quantitative data method was used. A Statistical Package for the Social Science (SPSS) was used for statistical analysis to compare views and opinions across groups. Qualitative data gathered from the nine key informants were analysed by use of qualitative content analysis (Flick, 2011; Urquhart, 2013). Information gathered from face-to-face interviews and casual discussions from staff (including tour guides) were transcribed verbatim. The information was put into

Table 7.1 Demographic profile of respondents (n = 50)

Gender	Frequency	Percentage (%)
Male	9	18
Female	41	82
Total	**50**	**100**
Age	**Frequency**	**Percentage (%)**
20–25	8	16
26–30	14	28
31–35	2	4
36–40	17	34
41+	9	18
Total	**50**	**100**
Education	**Frequency**	**Percentage (%)**
No schooling	6	12
Primary school	11	22
Junior secondary	14	28
Senior secondary	15	30
Tertiary/technical college	4	8
Total	**50**	**100**

categories, and a coding unit was defined to enable the data to be put into different levels of analysis (e.g., open coding) (Urquhart, 2013, p. 78–88). Field visits data and other material from participatory observation were collated, and field notes were organised into scripts as recommended by Groves et al. (2004).

7.5 Community Level Benefits of Cultural Tourism in Goo-Moremi

7.5.1 CBO/Trust and Joint Venture Partnership

Key informant interview analysis revealed that the Goo Moremi Mannonnye Development Trust (the Trust) began a JVP contract with the private operator in March 2013. The operator changed the name from Goo-Moremi Gorge to Goo-Moremi Resort, which could have been for marketing and promotional purposes. When asked if there has been any improved benefit (Table 7.2), the other key informant indicated that the JVP with the Trust was initiated to help safeguard the monument and at the same time create jobs for the community. He praised the initiative and said that the employees derive benefits as they are taught how to follow trails, cultural traditions, and tour guiding. Additionally, the household data analysis shows that the majority (88%) indicated that they benefit from the Joint Venture Partnership (JVP) of the Moremi community CBO/Trust and the private operator.

Table 7.2 Views on community benefits from CBO/Trust Tourism and JVP

Questions	Options	Frequency	Percent
Views on tourism benefits from CBO/Trust	Yes, very much	6	12.0
	Yes	38	76.0
	No	6	12.0
	Total	50	100.0

Only 12% of the household representatives pointed out that the community did not derive benefits. Overall, the research established that the CBO/Trust generate benefits to the community.

Following the question on the benefits, the respondents were asked to share information about the types of benefits they have received from their CBO/Trust and the private operator. Over 90% of the respondents said that they benefited through job opportunities, income generation, skill acquisition and improved life. About 11 people were employed as housekeepers, security guards and tour guides. Both the household heads and key informants stated that the majority of those employed came from within the community, while only two employees, the manager and accountant, were not residents of Moremi. As regards monetary benefits from doing business with the tourism operator, the key informant interview revealed that the community got P100, 000 (US$10000.00) annually as per the contractual agreement with the private operator.

Other intangible benefits highlighted from the JVP included free access (visits) of the residents to the Gorge. Similarly, the informants, for example, village elders, including headmen (*dikgosi*), Village Development Committee (VDC) members, pastors (*baruti*) and others, were allowed access to the site at no entry fee. As stated by one of the key informants, "*members of the community (villagers) are allowed to practice their cultural and belief systems at the gorge despite the lease agreement 'with the private operator'*." Cultural traditions included but were not limited to dancing and praying to gods (*badimo*), healing activities (dipping sick individuals into the sacred waters in the pools) and consultations with the gods for peace and protection. However, the private operator was tasked to ensure that the Gorge was kept safe and clean. The key informants alluded to the Gorge area having been kept clean since the engagement of the private operator.

The study has generally brought to the fore the question of improved living standards of the residents of Moremi. In response, the sizeable number of households (69%) indicated that the CBO/Trust had improved their living standards through job opportunities as some family members work as security guards, tour guides, cleaners (campsites and chalet) and others are engaged as labourers or do once-off jobs whenever there is a need. Monetary benefits (income) were also mentioned, while other benefits were visibility of the village (image), pride and happiness as local people associate with The Gorge even more (see Tomaselli, 2012). A smaller number of residents (17%) indicated that the CBO/Trust had not improved their living standards. Overall, it can be concluded that most of the villagers were positive that the CBO/Trust brought change to their lives.

7.5.2 Community Involvement in the Upkeep of the CBO/Trust

Knowledge and awareness about the existence of CBO/Trust within the village

Overall, respondents were asked if they were aware or had knowledge about a CBO/Trust in their village. The respondents were given three options to choose from which are Yes, very much, Yes and No options (Table 7.3).

Table 7.3 shows that 92% of the household heads indicated that they knew of the existence of the Trust within their community (Yes combined with Yes, very much), while 8% pointed out that they did not know. The lack of awareness was a surprise since the Trust is supposedly made up of all members of the community in the village. Those who do not know maybe newcomers to the village or extremely not interested in the village development issues.

7.5.3 Level of Commitment by the Community to Their CBO/Trust

The level of commitment to the CBO/Trust was measured by asking the respondents if they participated in the activities of the Trust. The respondents were asked to rate their level of commitment based on 0% no commitment to 100% very committed. The results show that almost half of the household representatives (48%), said that they did not participate nor commit any of their time to the CBO/Trust activities or operations, while 32% of the respondents committed 50% of their time to the CBO/Trust. Ten per cent committed 75% of their time to serve in the operations and management of the CBO/Trust, and only 8% said they were committed 100% of their time.

A substantial number of the informants stated there is generally a lack of commitment from the community about their CBO/Trust. Said, "*we are trying by all to update the people (residents) about the Trust during Kgotla meetings, but they still are not willing to take part in the affairs of the Trust*". Furthermore, the respondents were asked about the attendance of meetings dealing with the CBO/Trust related activities and operations. For this question, 50% reported that they do attend the Trust meetings, whilst the other 50% reported that they do not attend such meetings.

Table 7.3 Awareness of the existence of the CBO/Trust in Goo-Moremi

Question	Views	Frequency	Per cent
Awareness of the existence of the CBO/trust in the community	Yes, very much	3	6.0
	Yes	43	86.0
	No	4	8.0
	Total	50	100.0

For this, the key informants observed that many individuals do not turn up for CBO/ Trust meetings, and the village headman also cited lack of attendance as one of the problems caused by people's poor attendance of the general community Kgotla meetings.

Respondents were asked whether they can freely contribute their views and opinions towards decision making about the CBO/Trust during meetings in their community. Fifty-two per cent of the respondents indicated that they have had opportunities to contribute towards decision making about the Trust at *Kgotla* meetings. However, 34% said that they do take part in the activities leading to decision making about the Trust at *Kgotla* meetings, while 14% of them said that they sometimes make contributions.

7.5.4 Knowledge About Finance-Profits of the CBO/Trust

The majority (87.8%) of the household representatives said that they did not know much about the income generated or profits accrued from the CBO/Trust operations with the private operator. A nominal number (12.2%) of the respondents indicated knowledge and awareness about the financial situation of the CBO/Trust. The finding could be associated with either lack of consultation between those in the management of the joint venture contract and the operator since in response to an earlier question. This study showed that a significant majority were aware of the Trust and benefited from it. This could show that awareness of the existence of the Trust does not equate to a situation where all know its detailed financial operations. One expects that almost all the residents will be aware of the JVP, and the village benefits from it generally. While the financial status is disclosed in detail at the JVP Annual General Meeting, those who do not have financial literacy may not recall much of the shared information, and those who are not interested in the matters of the village may also fall within the 87.8%.

7.6 Challenges of the Joined Venture Partnership with the Community

7.6.1 Challenges Experienced Before the Establishment of the JVP with the CBO/Trust

The study findings indicated two main challenges that were experienced by the community prior to the development of the CBO/Trust and the joint venture with the private operator. Ninety-two percent of the respondents indicated that there were two problems in the village before the development of the CBO/Trust. The majority (86%) stated unemployment, whilst a few (6%) of respondents reported unsustainable use of natural resources as a problem. The rest, 8%, said that no challenges were experienced before the establishment of the CBO/Trust and the JVP.

The key informants observed that the biggest obstacle to the development of tourism in Goo-Moremi was poor communication infrastructure such as telephone facilities and the bad road to the Gorge site. They indicated that the telephone that was used by the Trust for bookings was placed at the village Kgotla (Traditional meeting place) some distance from the Gorge, as a result sometimes when clients and tourists call it was difficult to give truthful and up to date information about the gorge due to the distance from the village. The other challenge cited was the lack of electricity for lighting and heating at the Gorge. Thus, all these hindered development of the CBO/Trust tourism before the JVP since the community through their Trust could not attract overnight tourists who could stay in their lodging facilities (thatch chalets.

7.6.2   Poor Communication Between the Operator, CBO/Trust & Residents

Seventy-eight per cent of the respondents cited poor communication between JVP and the community regarding the plan of action, including new developments and tourism activities that interfered with the sacred nature of the site. About 14% of the respondents assertively indicated that development challenges greatly affect the CBO/Trust. However, a small proportion (8%) highlighted no problems or challenges regarding the operations of the CBO/Trust.

Almost all the key informants reacted with one voice to management issues with the private operator. In all, they mentioned personality clashes, possible ambiguity in role definitions, and a fear by the community that the engaged private operator could overrun their aspirations. One of the key informants representing employees remarked:

That manager (private operator) is rude, and want to be in control of everything, insulting us and disrespecting us as if we did not deserve respect.

7.6.3   Management Issues

Further, many respondents (76%) said that they were not involved in management, decision making, and finances of the JVP and the CBO/Trust. The respondents decried lack of transparency and accountability as the operator acted as the sole owner of the initiative. As perceived by some respondents, the operator appeared to have had the greatest power and the sole control of all activities, hence displayed no adherence to the initial business agreement of the JVP. This study did not conclusively establish what terms of the initial or current, business agreement was violated. Neither could it be established how serious the perceived violations were. It is assumed that some residents, who possibly may not even have an idea of the terms of agreements, may have their own ideas of what the agreement should have entailed. Such would then develop a view that some of the terms have not been adhered to. Subsequently, if the board members were tasked to ensure that the JVP terms were observed, it would have been unlikely to allow the terms of the agreement to be blatantly dishonoured. On the other hand, the partner would not be inclined to disrespect any stated JVP terms because that would be tantamount to killing the goose that lays the golden egg for their company.

The alleged incident where tourists either washed or swam in some pools at the Gorge, despite the understanding that such should not happen because the pools are sacred (see Mbaiwa, 2011; Njadingwe, 2015), may require stiff regulatory sanctions of the visitors' behaviour. It is unlikely to be a valid example to prove wilful disregard to the terms of the JVP deal. However, there must be vigilant tour guides who will prevent tourists' activities that are not acceptable.

7.7   Conclusions

The study revealed that community-based cultural tourism activities in Goo-Moremi village and the surrounding villages are relatively new. Tourism associated infrastructure and facilities, such as cultural lodges, traditional villages and open-air museum outlets, are limited or almost non-existent in some areas. However, there has been a steady increase in tourist numbers to the key attractions in the village, namely the Goo Moremi Gorge. There are some recent developments inside the gazetted Goo-Moremi Manonnye Conservation area, which include campsites with tourists' ablutions, thatched chalets, and gated entrance to the Gorge site.

Past studies on joint venture partnership between communities and the private operator in tourism-related projects have revealed mixed results (Kirtsoglou & Theodossopoulos, 2004; Mbaiwa & Sakuze, 2009; Lenao, 2017; Moswete & Thapa, 2016; see Movono & Dahles, 2017; Snyman, 2014). In this study, residents are beginning to reap benefits from community tourism and the joint partnerships in managing cultural heritage resources in Goo-Moremi. The benefits include, but are not limited to, the creation of employment opportunities, income generation, and heightened knowledge of conservation and protection of local heritage sites and resources. Subsequently, there is a sign of empowerment of those employed as guides at the Gorge as they have gained skills of tour guiding and handling and communicating with visitors (See Lenao, 2017; Movono & Dahles, 2017). Others from the CBO/Trust have been sent for short term training, for instance, on the front office, book-keeping and tour guiding. Thus, the study concluded that there is a sign of the social exchange theory (see Cook & Rice, 2006) in community-based cultural tourism in Goo-Moremi as both the residents, the private operator and resources have begun to benefit.

There was high awareness about the existence of the community CBO/Trust, yet there was limited knowledge about the finances and the general management of the Trust. Hence, the need to find ways to increase knowledge or raise awareness of the community Trust about the operations of the JVP. Other challenges outlined include a lack of commitment by the residents in the general affairs of the Trust, a low level of involvement in the general management and operations of the Trusts, and a tendency for the private operator to display the greatest power or influence in the management decision making processes. Poor communication regarding decision making and some activities at the Gorge were mentioned as key challenges. It can be concluded that in JVP tourism operations at a sacred site or monument, there is always scope for conflict between the proper use versus the desecration of a site (see Kirtsoglou & Theodossopoulos, 2004). This brings up the challenge of managing tourism in areas where there are spiritual observations. Thus, we recommend that cultural heritage tourism within local communities needs to be managed sensitively and responsibly, and care must be taken to monitor the scale and nature of development before the heritage sites become irrevocably damaged. Therefore, JVP contracts with communities and the private operator must be adhered to with mutual respect and trust if tourism is to benefit the resources, JVP business partner and the residents.

References

Adams, G. F., & Wyckoff, J. (1971). *Landforms*. Golden Press.

Bayliss, T. (1995). *A concise advanced geography*. Oxford University Press.

Bernard, H. R. (2000). *Social research methods: Qualitative and quantitative approaches*. Sage.

Blackstock, K. (2005). A critical look at community-based tourism. *Community Development, 40*(1), 39–49.

Botswana Tourism Development Program (BTDP). (2003). *Botswana national ecotourism strategy*. Gaborone.

Cassidy, L. (2000). CBNRM and legal rights to resource in Botswana. In *CBNRM Support Programme Occasional Paper N0.4*. IUCN/SNV.

Cave, C., & Negussie, E. (2017). *World heritage conservation: The world heritage convention, linking culture and nature for sustainable development*. Routledge.

Child, B., Ward, S., & Tavengwa, T. (1997). Zimbabwe's CAMPFIRE Programme: Natural resource management by people. World Conservation Union – IUCNROSA. *Environmental Issues Series, 2*.

Clark, A. N. (1998). *The penguin dictionary of geography* (2nd ed.). The Penguin Books.

Cook, K. S., & Rice, E. (2006). Social exchange theory. In J. Delamater (Ed.), *A handbook of sociology and social research* (pp. 53–76). Kluwer Academic /Plenum Publishers.

Creswell, J. W., & Clark, V. L. P. (2011). *Designing and conducting mixed research*. Sage.

Department of Wildlife and Natural Parks. (1999). *Joint venture guidelines: A guide to developing natural resources based business ventures in community areas*. Department of Wildlife and National Parks.

Dichaba, T. (2009). *From monuments to cultural landscapes: Rethinking heritage management in Botswana*, Unpublished MA Dissertation, Rice University, USA.

Eagles, P. F. J., McCool, S. F., & Haynes, C. D. (2002). *Sustainable tourism in protected areas: Guidelines for planning and management. World Commission on protected areas (WCPA). Best practice protected area guidelines series N0.8.* The World Conservation Union.

Flick, U. (2011). *Introducing research methodology: A beginner's guide to doing a research project.* Sage.

Geoflux Pty. (2009). *Environmental impact assessment for Goo-Moremi conservation area.* Unpublished Scoping Report. Geoflux.

Government of Botswana (GoB). (1990). *Tourism policy.* Government paper No. 2 of 1990z. Government Printers.

Government of Botswana (GoB). (2001). *The Botswana National Atlas.* Department of Surveys and Mapping. Government Printer.

Government of Botswana (GoB). (2007). *Natural resources management policy. Government Paper N0. 2 of 2007.* Ministry of Environment, Wildlife and Tourism. Government Printer.

Groves, R. M., Fowler, F. J., Jr., Couper, M. P., Lepkowski, J. M., Singer, E., & Tourangeau, R. (2004). *Survey methodology.* Wiley.

Hannam, K., & Knox, D. (2010). *Understanding tourism: A critical introduction.* Sage.

Honey, M. (1999). *Ecotourism and sustainable development: Who owns paradise?* Island Press.

Howard, P. (2003). *Heritage: Management, interpretation, identity.* Continuum.

Keitumetse, S. O. (2016). *African cultural heritage conservation and management: Theory and practice from southern Africa.* Springer International.

Keitumetse, S. O., & Nthoi, O. (2009). Investigating the impact of world heritage site tourism on the intangible heritage of a community: Tsodilo Hills World Heritage site, Botswana. *International Journal of Intangible Heritage, 4,* 144–151.

Kirtsoglou, E., & Theodossopoulos, D. (2004). They are taking away our culture: Tourism and culture commodification in the Garifuna Community of Roatan. *Critique of Anthropology, 24*(2), 135–157.

Lavraks, P. (2008). *Encyclopedia of survey research methods:* Online ISBN. Retrieved from: http://srmo.sagepub.com/view/encyclopedia-of-surveyresearch -methods/nso3.xml

Lenao, M. (2017). Community, state and power-relations in community-based tourism on Lekhubu Island, Botswana. *Tourism Geographies, 19*(3), 483–501. https://doi.org/10.1080/1461668 8.2017.1292309

Lowenthal, D. (2006). Natural and cultural heritage. *International Journal of Heritage Studies, 11*(1), 81–92.

Manwa, H., Moswete, N., & Saarinen, J. (Eds.). (2016). *Cultural tourism in southern Africa.* Channel View.

Mbaiwa, J. (2005). Enclave tourism and its socio-economic impacts in the Okavango Delta, Botswana. *Tourism Management, 26*(2), 157–172.

Mbaiwa, J. (2011). Cultural commodification and tourism: Goo-Moremi community. Central Botswana. *Royal Dutch Geographical Society, 102*(3), 290–301.

Mbaiwa, J. (2013). *Community-based natural resources management (CBNRM) in Botswana.* Unpublished final CBNRM status report of 2011–2012. Kalahari Conservation Society Secretariat.

Mbaiwa, J. E., & Sakuze, L. K. (2009). Cultural tourism and livelihood diversification: The case of Gcwihaba Caves and XaiXai Village in the Okavango Delta, Botswana. *Journal of Tourism and Cultural Change, 7,* 61–75.

Mbaiwa, J. E., & Siphambe, G. (2019). Building cultural resilience in community based tourism. In M. Mkono (Ed.), *Positive tourism in Africa.* Routledge.

Mbaiwa, J. E., & Hambira, W. L. (2020). Enclaves and shadow state tourism in the Okavango Delta, Botswana. *South African Geographical Journal, 102*(1), 1–21.

Mmopelwa, G., & Mackenzie, L. (2020). Economic assessment of tourism-based livelihoods for sustainable developments: A case of handicrafts in Southern and Eastern Africa. In M. T. Stone, M. Lenao, & N. Moswete (Eds.), *Natural resources, tourism and community livelihoods in Southern Africa: Challenges for sustainable development* (pp. 235–253). Routledge.

Moswete, N., & Lacey, G. (2014). Women cannot lead: Empowering women through cultural tourism in Botswana. *Journal of Sustainable Tourism, 23*(4), 600–617.

Moswete, N., Thapa, B., & Lacey, G. (2009). Village-based tourism and community participation: A case study of Matsheng villages in Southwest Botswana. In J. Saarinen, F. Becker, H. Manwa, & D. Wilson (Eds.), *Sustainable tourism in Southern Africa: Local communities and natural resources in transition* (pp. 189–209). Clevedon.

Moswete, N., & Thapa, B. (2016). An assessment of community-based ecotourism impacts: A case study of the San/Basarwa communities of the Kalahari, Botswana. In K. Iankova, A. Hassan, & R. L'Abbe (Eds.), *Indigenous people and economic development: An international perspective* (pp. 223–237). Routledge.

Moswete, N., & Thapa, B. (2018). Local communities, CBOs/trusts, and people–park relationships: A case study of the Kgalagadi Transfrontier Park, Botswana. *The George Wright Forum, 35*(1), 96–108.

Movono, A., & Dahles, H. (2017). Female empowerment and tourism: A focus on businesses in a Fujian village. *Asia Pacific Journal of Tourism Research, 22*(6), 681–692.

Ndubano, E. (2000). *The economic impacts of tourism on the local people: The case of Maun in the Ngamiland sub-district, Botswana.* [Unpublished M.Sc. Thesis,] Department of Environmental Science, University of Botswana, Gaborone.

Njadingwe, K. (2015). Diversifying Botswana's tourism product: The potential of goo-Moremi village as a cultural-heritage attraction site. Unpublished project. In *Department of Tourism and Hospitality Management* (p. 42). University of Botswana.

Okech, R. N. (2007). Empowering women through involvement in ecotourism: reflections on Africa. In M. E. Kloek & R. van der Duim (Eds.), *Local communities and participation in African tourism* (pp. 107–117). *Thematic proceedings of ATLAS Africa conferences.*

Rossman, G. B., & Rallis, S. F. (2017). *An introduction to qualitative research: Learning in the field* (4th ed.). Sage.

Rozemeijer, N. (2001). *Community-based tourism in Botswana. Case studies in Xai-Xai, D'Kar and Khwai.* Unpublished SNV Publications.

Saarinen, J. (2010). Local tourism awareness: Community views on tourism and its impacts in Katutura and king Nehale conservancy, Namibia. *Development Southern Africa, 27*(5), 713–724.

Saarinen, J. (2012). Tourism, indigenous people, and the challenges of development: The representations of Ovahimbas in tourism promotion and community perceptions toward tourism. *Tourism Analysis, 16*(1), 31–42.

Saarinen, J., Moswete, N., & Monare, M. J. (2014). Cultural tourism: New opportunities for diversifying the tourism industry in Botswana. *Bulletin of Geography. Socio-economic Series, 26*, 7–18.

Saarinen, J., & Rogerson, C. M. (2015). Setting cultural tourism in southern Africa. *Nordic Journal of African Studies, 24*(3&4), 207–220.

Salazar, N. B. (2012). Community-based cultural tourism: Issues, threats, and opportunities. *Journal of Sustainable Tourism, 20*(1), 9–22.

Sebele, L. S. (2010). Community-based tourism ventures, benefits, and challenges: Khama rhino sanctuary trust, central district, Botswana. *Tourism Management, 31*(1), 136–146.

Siphambwe, G., Mbaiwa, J., & Pansiri, J. (2017). Cultural landscapes and tourism development in Botswana: The case study of Moremi Gorge in Eastern Botswana. *Botswana Journal of Business, 10*(1), 117–137.

Snyman, S. (2012a). The role of tourism employment in poverty reduction and community perceptions of conservation and tourism in southern Africa. *The Journal of Sustainable Tourism, 20*(3), 395–416.

Snyman, S. (2012b). Ecotourism joint ventures between the private sector and communities: An updated analysis of the Torra Conservancy and Damaraland Camp partnership, Namibia. *Tourism Management Perspectives, 4*, 127–135.

Snyman, S. (2014). Partnership between a private sector ecotourism operator and a local community in the Okavango Delta, Botswana: The case of the Okavango Community Trust and Wilderness Safaris. *Journal of Ecotourism, 13*(2–3), 110–127.

Spenceley, A., & Goodwin, H. (2007). Nature-based tourism and poverty alleviation: Impacts of private sector and parastatal enterprises in and around Kruger National Park, South Africa. *Current Issues in Tourism, 10*(2–3), 255–277.

Statistics Botswana. (2016). *Tourism statistics annual report 2015. Publication catalogue 2016.* https://www.statsbots.org.bw/tourism. Accessed 21 Mar.

Stevens, P. W., & Jansen, R. (2002). *Botswana national ecotourism strategy.* Unpublished Final Report. Government of Botswana.

Stone, M. T. (2015). Community-based ecotourism: A collaborative partnerships perspective. *Journal of Ecotourism, 14*(2–3), 166–184.

Tomaselli, K. (2012). *Cultural tourism and identity: Rethinking indigeneity.* Brill.

UNWTO. (2018). UNWTO tourism highlights, 2018 edition. e-unwto.org/doi/pdf/10-1811/9789284419876. Accessed April 2022.

Urquhart, C. (2013). *Grounded theory for qualitative research: A practical guide.* Sage.

Van der Jagt, C. J., & Gujadhur, T. (2002). Practical guide for facilitating CBNRM in Botswana. *CBNRM support programme. Occasional paper No. 8.* Botswana SNV/IUCN.

Van Vuuren M. J. (2019). Tourism report: *Africa's Tourism Market for June 2019.* Retrieved from https://www.bizcommunity.africa/article/410/373/192385.htm. 1 Mar 2021..

Veal, A. J. (2006). *Research methods for leisure and tourism: A practical guide (3rded).* Pearson Educational Limited.

White, R. (2000). *Tourism development plan for the Moremi Gorge.* Moremi Manonnye Development Trust Unpublished Draft Final Report.

Weaver, D. (2001). *Encyclopedia of ecotourism.* Wiley.

White, R. (2001). *Integrated development and management plan for the Moremi Gorge.* Unpublished Final Draft. Moremi Manonnye Development Trust Report.

WTTC (2017). Travel and tourism economic impact, march 2017 forecast. http://www.org. Accessed March 2022.

WTTC (2019). The economic imapct reports. Regional Overview. wttc.org/research/economicimpactss. Accessed April 2022.

Bontle Elijah holds a B. in Environmental Science from the Department of Environmental science, University of Botswana. She also holds an Occupational Health and Safety certificate from the Ministry of Health. In 2014 she was hired by the Tonota Sub land board as a land registration intern officer and then moved to work for the Nata Land Board as the Land Adjudication Overseer since 2018. Her research interest revolves around environmental issues, community development, Land issues, and cultural tourism.

Naomi N. Moswete is a Senior Lecturer in the Department of Environmental Science, University of Botswana. Her research interests include Human geography, tourism as a strategy for rural development, community-based tourism; Transboundary conservation areas –ecotourism nexus, parks–people relationships, heritage management & cultural tourism.

Masego A. Mpotokwane (PhD), is a Human Geography lecturer, the University of Botswana in the Department of Environmental Science. He has researched and published on land use issues, sustainable livelihoods, tourism development, disaster risk reduction and imp acts of human use on the land. His research interest includes sustainable development and human impacts on the physical environment.

Chapter 8
Socio-economic Impacts of Community-Based Ecotourism on Rural Livelihoods: A Case Study of Khawa Village in the Kalahari Region, Botswana

Naomi N. Moswete, Jarkko Saarinen, and Brijesh Thapa

8.1 Introduction

Since the 1990s, ecotourism has emerged as a form of tourism that emphasises the ideals of nature conservation and community participation and benefits in tourism development (Cobbinah et al., 2017; Stone & Nyaupane, 2015; Makwindi & Ndlovu, 2021). Emphasis is on rural communities that subsist on natural resources found in abundance within their areas of abode. In general, ecotourism promotes tourism activities that are both nature-based and cultural in character, and its principles are based on sustainable tourism (Anup et al., 2020; Wood, 2017), responsible tourism (Manning et al., 2017) or community-based tourism (Lorio & Corsale, 2014; Mmopelwa & Mackenzie, 2020; UNWTO, 2018). In some studies, ecotourism has been characterised and contextualised as nature-based tourism (Fennell, 2015; Borges de Lima & Green, 2017) with links to tourism in protected areas such as national parks and game and nature reserves (Dhakal & Thapa, 2015; Garekae et al., 2020; Stone & Nyaupane, 2015).

N. N. Moswete (✉)
Department of Environmental Science, University of Botswana, Gaborone, Botswana
e-mail: moatshen@ub.ac.bw

J. Saarinen
Geography Research Unit, University of Oulu, Oulu, Finland

School of Tourism and Hospitality, University of Johannesburg, Johannesburg, South Africa
e-mail: jarkko.saarinen@oulu.fi

B. Thapa
School of Hospitality & Tourism Management, Oklahoma State University,
Stillwater, OK, USA
e-mail: bthapa@okstate.edu

© The Author(s), under exclusive license to Springer Nature
Switzerland AG 2022
J. Saarinen et al. (eds.), *Southern African Perspectives on Sustainable Tourism Management*, Geographies of Tourism and Global Change,
https://doi.org/10.1007/978-3-030-99435-8_8

Although there has been a strong focus on natural environments in ecotourism research, Cater (1993) has posited that culture must be recognised as an important aspect of ecotourism. This highlight the idea that natural and cultural landscapes should not be separated. Accordingly, Caldwell (1996) further contends that most landscapes we consider as natural have some cultural influences and human impacts based on their historical evolution (Diallo & Proulx, 2016; Mulder & Coppolillo, 2005; Pribudi, 2020). In this respect, ecotourism has the potential to protect and benefit both natural and cultural environments (Cobbinah et al., 2017). However, it is important to note that there may also be environmental and social challenges in ecotourism development (Anup et al., 2020; Moswete & Mavondo, 2003; Saarinen & Manwa, 2008). Increased tourism based on conservation areas and local community resources has sometimes led to crowding, conflicts and over utilisation of natural resources, which has created serious threats to environment and/or community livelihoods (see Duim van der et al., 2011; Panta & Thapa, 2017). Indeed, uncontrolled tourism development can be destructive since it can enable nature enthusiasts to penetrate further afield, exploring natural areas which could not otherwise have been accessed, thus, exposing them to tourism-related pressures and damages (Fennell, 2015). Furthermore, some unsustainable practices of local ecotourism have been observed in which enclave tourism prices out local people from participation (Anderson, 2011; Mbaiwa, 2005; Mbaiwa & Hambira, 2020; Saarinen, 2010, 2017; Saarinen & Wall-Reinius, 2019; Scheyvens & Biddulph, 2018).

Despite potential challenges thereof, ecotourism has become a tool for economic development and environmental protection in many developing nations, including Botswana (Mbaiwa, 2008, 2015; Saarinen et al., 2020; Stone & Nyaupane, 2015). If it is to be beneficial and sustainable, local communities should be allowed to derive a substantial amount of the socio-economic benefits generated by the industry (Fennell, 2015; Moswete et al., 2020). Similarly, community-based ecotourism has been identified as an alternative option for economic growth in Botswana (Garekae et al., 2020; GoB, 2007). Based on these premises, this study is interested in examining the impacts of community-based ecotourism development in Khawa in southern Kgalagadi, in the Kalahari desert region of western Botswana, which is a small, marginalised and remote village. The chapter aims at assessing the socio-economic impacts of community-based ecotourism (CBE) on the livelihood of the residents. The objective of the study is to assess the socio-economic impacts of community-based ecotourism in Khawa and to examine the perceived impacts of tourism on the environment at the area. Furthermore, the study seeks answers for the following specific sub-questions: (a) has ecotourism benefitted local residents of Khawa?; (b) How has tourism impacted the local landscape at Khawa; and (c) Has community-based ecotourism empowered the local residents of Khawa?

8.2 Community-Based Ecotourism

Based on the global tourism trends and developed policies in southern Africa (see Monare et al., 2016; UNWTO, 2018), there has been an increasing interest in community-based ecotourism (Stone & Stone, 2020; Moswete & Thapa, 2018). In practice, CBE is organised in various ways in different socio-economic and political contexts. In general, the key principles of CBE related to the requirements to involve so called gateway communities and villagers living adjacent to protected areas to enhance their income (Anup et al., 2020; Kavita & Saarinen, 2016; Panta & Thapa, 2017; Moswete et al., 2012; World Bank Group, 2021). CBE may also help rural residents to refrain from agricultural dependence as local people who participate in CBE can receive benefits (income, employment) from tourist consumption (Chiutsi & Saarinen, 2017, 2019; Dhakal & Thapa, 2015; see Pribudi, 2020). Well managed CBE can also restore degraded rangelands, revive cultures, protect and preserve endangered species of fauna and flora, reduce resource conflicts, and improve the living standards of rural communities (Greeffe, 2009; Moswete et al., 2009; Mbaiwa, 2008). Indeed, in many cases in southern Africa, CBE has contributed positively to communities (Kimaro & Saarinen, 2020; Makwindi & Ndlovu, 2021; Mbaiwa, 2013; Mmopelwa & Mackenzie, 2020; Moswete et al., 2012).

In Botswana, some working examples are the Nqwaa Khobee Xeya Development Trust (Kgalagadi north) and Sankoyu Tshwaraganyo Development Trust (northern Botswana) (Arntzen et al., 2003; Mbaiwa, 2013; Moswete & Thapa, 2018). Concerning Botswana's Ecotourism Strategy (GoB, 2003), CBE implies that a community is "caring for its natural resources in order to gain income through tourism and is using that income to better the lives of its people; it involves conservation, business enterprises, and community development" (GoB, 2007).

In positive cases, CBE commonly involves business initiatives that are wholly owned by local communities and are inherently less dependent on foreign suppliers. The small-scale community initiatives are communally owned, managed and operated, and the benefits accrue to the residents (Greeffe, 2009). Furthermore, in successful cases of CBE, local people and all villagers are trained on issues of how to initiate, start and operate a CBO/Trust (GoB, 2007; Moswete & Thapa, 2018). Further, communities are supported and encouraged to venture into CBE initiatives by forming community-based organisations (CBOs) or Trusts (Kimaro & Saarinen, 2020; Mbaiwa, 2013; Moswete & Thapa, 2018). For Botswana, the CBOs or Trusts are tourism initiatives that are owned by one or more communities, or run as joint venture partnerships with the private sector with equitable community participation, as a means of using natural and cultural resources in a sustainable manner to improve local livelihoods and safeguard the environment (GoB, 2007; Mbaiwa, 2013).

However, as previously noted, there can be challenges and failures involved with CBE due to varying resource availability as well as management plans and policies (Moswete et al., 2020; Stone & Stone, 2020). Based on previous research, some local communities have not benefited from tourism activities due to lack of skills and various capacities about the industry (Moswete & Lacey, 2014; Saarinen, 2010).

Some studies have revealed that tourism can only benefit the non-local tour operators, while the residents were found to have had limited knowledge to facilitate
ecotourism ventures (Cobbinah et al., 2017; Mbaiwa, 2003).

8.3   The Case Study: Socio-economic Impacts of Community-Based Ecotourism in Khawa

8.3.1   Khawa Village

Khawa is a village in the Kalahari region of southwestern Botswana. Khawa is one
of the smallest villages in the Kgalagadi District, with a population of 817 in 2011
(Central Statistics Office, 2013). The village is growing, and it has steadily become
a so-called complete village (see the revised Botswana Settlement Strategy of 1998)
with one primary school, mobile health stop, Kgotla (traditional meeting place)
offices and the village development quarters (Moswete, 2009). The roads connecting Khawa village with other nearby settlements, villages and farm areas are dirt
roads, treks and pathways (Moswete & Thapa, 2015). The ethnic groupings of
Khawa include Bangologa, Batlharo, Coloureds and Nama and they live under the
leadership of Kgosi Titus Manyoro. The history of Khawa shows that the people of
this village hailed from South Africa and as such many of their relatives reside in
South Africa. The livelihood activities for the community are subsistence arable
farming and rearing of small stock (goats, sheep). A few individuals in the community own cattle, donkeys, and horses for subsistence purposes. However, poor and
infertile sandy soils make crop farming a problem, while low to unreliable rainfall,
water shortage and poor pastures (desert) render keeping of livestock (cattle, donkeys, goats) a challenge to residents (Moswete, 2009) (Fig. 8.1).

Geographically, the Khawa village is situated about 20 kilometres from the
boundary of the southern part of the Kgalagadi Transfrontier Park (KTP), which
straddles between Botswana and South Africa. Khawa is one of the very few villages found close to the Park's boundary line. Historically, the Park was made up of
two national parks: Gemsbok National Park in Botswana and the adjoining South
African Kalahari Gemsbok National Park (Southern Africa Development
Community (SADC), 2020). In 1999, the two parks were merged and became
Africa's first officially declared transboundary (peace) park. The Park is still largely
the only open peace park where tourists can move freely across the international
boundaries of the park. Tourists who visit the park pay entrance fees, and according
to the shared management and operation of the KTP, entrance fees are shared
equally between Botswana and South Africa. However, each country is responsible
for developing its tourism-related facilities.

The village of Khawa is located within the Wildlife Management Area (WMA)
known as KD 15 (Mbaiwa, 2013). The specific WMA is a buffer zone with links to
the Kgalagadi Transfrontier Park. In general, land uses allowed in WMAs arc classified in two – consumptive or non-consumptive wildlife utilisation. With respect to

Fig. 8.1 Queueing to collect water that is supplied through a water bowser to Khawa village Water supply shortages are common in the Kalahari region. (Photo: J. Saarinen)

tourism development, the focus in WMAs is on nature-based tourism activities. In the case of the Khawa village, it was an unknown small village with very little or no tourism taking place there till the year 2000. However, with the advent of the wild-life conservation policy of 1986, the Botswana National Parks Act of 1992 and the community based natural resources management policy of 2007, wildlife hunting became tightly controlled. At the same time, poaching or illegal hunting of wild animals and birds became problematic (Moswete et al., 2012). As a partial response, the Government encouraged rural communities to venture into tourism through the introduction of the Community Based Natural Resources Management (CBNRM) (Mbaiwa, 2013). As a result, the Khawa village became one of the first in the Kgalagadi district to establish a safari-based tourism CBO-Trust (BTDP, 2000; Moswete et al., 2009). It is formally known as Khawa Kopanelo Community Development Trust (Arntzen et al., 2003; Mbaiwa, 2013; Moswete et al., 2012).

Based on the previous studies, tourism has had limited benefits for the members of the community of Khawa (Moswete, 2009). Community benefits derived from the safari hunting included meat, part-time jobs (e.g., cooks, trekkers, skinners, and campsite caretakers), income and overall pride for being associated with the KTP (Arntzen et al., 2003; Mbaiwa, 2013; Moswete et al., 2012). Reduced poaching in the area has been noted as local people began to understand the value of protecting wild animals in their area (Moswete, 2009). Also, the residents via their CBO/Trust began to freely become part of the community tourism project and got involved in conservation activities of wildlife in and around the village of Khawa (Moswete et al., 2012). Further, a CBO/Trust built a shelter as an office, purchased a truck – an off-road vehicle and people were assisted with some money and food during bereavement and ill-health (Moswete, 2009; Moswete et al., 2012). In 2012, a government initiative popularly known as the Khawa dune challenge-tourism related annual event was introduced, and Khawa village was identified as a place suitable for the event. At the initial stage, the community was not happy with the event as it appeared like it was imposed on them. However, like many tourism development initiatives, the introduction of annual event-based tourism in the community brought challenges but also potential opportunities.

8.3.2   Methods

Research materials were collected via a semi-structured systematic household interview; every other home or plot that intersects paths and roads was selected and visited for interviews. The head of the household was requested to participate, and a consent form was signed by the interviewee before each interview. In instances when the head of the household was not home, any member of the family who was 18 years or older and had lived in the village or district for at least 12 months was asked to participate. As a result, a total of 75 household heads (more than 50% of all 128 households) were systematically selected, while 15 key informants were selected through the purposive method. Unstructured casual discussions with the village headman, manager of the CBO/Trust and the Village Development Committee (VDC) secretary were conducted. An observation method with a situational analysis approach (see Koutra, 2010) was used during two of the Khawa Dune Challenge and Cultural Events to collect more information for this study.

Additional information and research materials were sought through a telephone interview from June to November 2020. Purposive and snowball sampling approaches were used to select key informants (see Patton, 1999; Robins, 1963; Veal, 2006). This was initiated by first identifying one of those who play a role in the village development activities of Khawa. It was through the initial contact that some names of persons were recommended, and telephone and mobile phone numbers were availed to the lead researcher. Once an individual was contacted by phone, a survey sheet was sent via a mobile phone application and electronic mail (see Holliday, 2001; Veal, 2006). In total, a 15-item collection survey sheet was sent out to identified key people, and ten were completed and returned.

8.3.3 Demographic Characteristics (Household Interview)

There were more female participants (59%) than males. Many were born (natives) of Khawa and 60 of them had lived in the village all their lives, while others had lived there for more than 10 years (24%). The remaining participants had resided in the village for less than 10 years (16%). The youngest participant was 18, and the oldest was 81 years, with about 83% of them being able to read and write. In the data the ethnicity of Khawa was comprised of Batlharo (69.3%), Coloreds (10.6%), Bakgalagadi (6.7%), Bangologa (5.3%), BaNama and Baherero with 1.5% respectively and others (5.3%). The estimated household monthly income among the 75 participants ranged from less than P500 (USD45) to P3500 (USD320). Almost 21% had a formal job, 20% were part-time, and unemployment stood at 32%. Many families (23%) were involved in the government welfare programs known as the drought relief project (namula leuba), whereas 15% were registered under the destitution program, where they receive food baskets monthly. Fewer members of the community (10.6%) were involved in farming (sheep, goats), while only three persons mentioned craft as a source of income.

With respect to the key informant telephone survey, there were slightly more females (6) than males (4); with ages ranging from 27 to 49 years old. Those who took part in the mobile phone interview came from 6 villages in Kgalagadi, with two from Gaborone and one was from Palapye. These representatives included a CBO/Trust officer, village elders, tourism development management offices, environmental and safety managers, teachers, and retirees). Nearly all of them had visited Khawa for purposes including an officer on duty, field visits, business, and as spectators (visitors/tourists) during the annual community event.

Since the study is exploratory, data was analysed by descriptive statistics where proportions and pivot tables were used to validate data and present the results. Qualitative data generated from open-ended questions and information obtained from the interview were transcribed and analysed (see Groves et al., 2004; Holliday, 2001; Miles et al., 1994). Observation information gathered (unobtrusive technique) was incorporated in the analysis (Rossman & Rallis, 2017; Veal, 2006). Thus, open, axial, and selective coding of qualitative data was used to derive key findings and conclusions (Strauss & Corbin, 1998; Urquhart, 2013).

8.3.4 Socio-economic Impacts: Economic-Tourism Related Activities

The community of Khawa owns a CBO/Trust known as Khawa Kopanelo Community Development Trust (KKDT). The community-initiated a joint venture safari hunting activity through their KKDT with the assistance of the CBNRM-CBO government initiative program. With respect to the interview, the respondents indicated that they had engaged a safari hunting operator to manage their tourism enterprises and activities for them and that they had a contractual agreement as stipulated in the joint venture legal contract document (see Thusanyo Le fatsheng, 2005). The CBO/Trust operated within the concession area known as the Kgalagadi District (KD) 15, community-based camping activities.

The KKDT offers several safari and ecotourism activities to local and international tourists, especially those with vast interest in nature, wilderness, wildlife and cultures and history. The safari operator offers camping services at designated fenced off camping grounds, which has a long drop shower, pit latrine, garbage bins, fireplace, and barbecue stands with some sitting areas (see Moswete et al., 2009). The socio-economic benefits revealed by both household representatives and almost all the key informants were casual, part-time and full-time employment, income (salary and tips), business opportunities and game meat. Empowerment activities for the villagers were through formal meetings (seminars, workshops), informal training (Kgotla meetings), and some benefitted through on job training (CBO/Trust and VDC members).

The community derive monetary benefits through the CBO/Trust and other ecotourism activities in the village. According to the interview with the CBO/Trust manager and the village headman, the safari operator pays the community a lump sum of money. The money is shared in which the CBO/Trust management subdivides cash amongst all households in the community. For instance, in the period 2011/2012 the Khawa CBO/Trust generated revenue amounting to P16, 000 from tourism activities (Mbaiwa, 2013) and this was shared amongst individual homes.

Still, on benefits at the community level, one of the key informants observed:

> From the 2019 Khawa Dune Challenge and cultural tourism event socio-economic survey report that the revenue accrued to the community of Khawa from [tourism-related businesses, . . . Do It Yourself (DIY) campsite and the resident vendor accrued [stoop up to] P130, 341.45 for just the Khawa annual event weekend.

Other individuals benefitted by receiving direct cash income. This was based on selling handcrafted ornaments (*wooden spoons, bone tools*), herbal teas (e.g., *mosukujane)* and wild berries (e.g., *moretlwa*), wild mushrooms (*mahupu*), for example. Compared to other communities or villages in Kgalagadi, only a small number of people were involved in craftwork such as skin tanning for mats/carpets, sofa cushions and handbags. Unlike in the rest of Kalahari, only an insignificant number of residents were involved in beadwork using ostrich eggshells in Khawa. From the household survey, a large proportion of the respondents (94.6%) said that CBE was essential to their community. However, when household respondents were asked if revenue from community-based tourism benefited many persons in the community, slightly more than a half (52%) of the household representatives answered yes to the statement, while quite a high number of the respondents (35%) said there were no benefits.

Furthermore, it was discovered that Khawa was endowed with geosites and geotourism resources for adventure and photographic tourism. This small and remote community has also become a popular village that hosts the annual Khawa dune challenge tourism activity in southern Kalahari. The village and its beautiful landscape have become an attraction in its own right (see Saarinen et al., 2012). In addition, and from the recent interview data, the key informants resonated that the community of Khawa benefited from the annual Khawa dune challenge and cultural event because a good number of them profit directly from the 'Do it yourself' (DIY) campsite accommodation within their homes, personal vendor stalls, sale of foods and handcrafted local souvenirs to visitors. As one key informant observed: *"Residents benefitted through their community Trust (CBO) business, and from renting out their undeveloped residential plots/land for camping".* Moreover, the CBO/Trust also obtains monetary benefits from tourism business in which they offer lodging and other services including food.

Subsequently, the findings revealed that tourism had opened opportunities for employment within the local community. So far, some members of the community have secured jobs from tourism-related businesses such as CBO/Trust guesthouse and campsite. Some residents work as part-timers; causal labourers, while others said that they were engaged to do menial jobs for example, mending of the Park (KTP) fence, campsite cleaners, watching/caretakers of campsites, skinners and local guides and others were engaged as interpreters during hunting expeditions. For instance, a sizeable number (65.3%) of them said the KTP provides jobs for people in their community, while a substantial number (89.3%) were happy that their village is situated closer to the KTP. This has opened opportunities for them to secure jobs at the Park whilst others were employed by the safari operator. However, we still found that unemployment at the time of the study was relatively high (32%) in Khawa, while self-employment was 24%; formal employment at 21.3% and part-time or casual jobs was 20% among the household respondents.

Similarly, the key informants revealed that there had been an increase in employment opportunities in the village, indicating that some of the residents get to work as casual labourers, picking litter and providing security services (policing/watching) towards unwarranted behaviours in the village and even poaching of wild animals. Other individuals from the village set up their own stalls and kiosk outlets to sell food and groceries creating jobs for helpers who are engaged from the village. In terms of tourism infrastructure and development, CBE facilities for tourism were minimal and standard during the first part of the study, but there were some developments: e.g., the CBO/Trust had purchased a 4x4 off-road vehicle with a tourism cash income obtained from the safari operator. The community benefited from the Trust/CBO as the vehicle formed a bigger part of the social capital (welfare support) for everyone in the village. For example, in the event of sickness or death in the community, the vehicle is used to ferry them to and from a nearby clinic in Middlepits village or to a hospital located 167 kilometres, quite a distance away in Tsabong village (the largest village in southern Kalahari).

The study also discovered that most of the respondents (89.3%) held the perception that they were happy to have their village adjacent to the Park (KTP). In a similar fashion, the key informants also echoed community attachment to the Park (KTP) as it was a home for their forefathers and revealed evidence of physical archaeological remains that include dwelling foundations, trough and potsherds at Rooipuits (Fig. 8.2).

Fig. 8.2 Dwelling remains and old boreholes found near Rooipuits campsite inside the Kgalagadi Transfrontier Park. (Photo: N. Moswete)

The developmental changes that we discovered at the study site fall as part of the Botswana government mandate to diversify tourism away from urban centres to rural areas to promote community tourism and improve lives. There were some related concerns with tourism development. At the top of their concern was an increase in incidents of HIV/AIDS infections and transmission (92%), followed by an increase in social ills (e.g., crime) (86.7%), and that tourism would change their cultural traditions (64%). However, they appeared to be slightly less concerned about the statement that tourism would destroy the environment (49.3%).

The key informants were also asked to share their views about community tourism development in Khawa. Specifically, they were asked to give out their views and opinions on maximum three major 'things' (issues) they disliked about tourism in Khawa. Nearly all of them stated that the gravel road that adjoins Khawa to other villages (main access road) was in a poor state of disrepair as it was damaged, with potholes and was too dusty due to increased traffic and overuse.

> they should consider developing and improving the road from Khuis village to Khawa so as to minimise the likelihood of accidents because the road as is, is very bad with potholes and the sand is loosened and is dusty as traffic to the village increases during the Khawa dune challenge tourism event.

In addition, the issue of noise pollution was highlighted. Some respondents felt that noise pollution disrupted village peace and tranquillity during the event. One of the respondents stated that some residents even visit relatives as they are not able to withstand what they refer to as 'commotion' during the event weekend. There was also a mention of crowding problems during the tourism event week. There were no tourism-related issues raised beyond the event. Thus, it seems that the tourism-related challenges are mainly linked to the Khawa Dune Challenge and Cultural

Event, which was considered as creating changes to the village. This also includes challenges in the economic system of the event. Since many tourism services are imported into the village for the event, such as camping equipment, tents, hired toilets, foodstuff, most of the generated revenue from tourism does not remain in the community but leak out to those operators who come from larger villages such as Tsabong, or towns such as Jwaneng and Gaborone.

Despite the challenging issues, almost all the key informants observed that there were some recent positive developments that have taken place within the community due to tourism. There was a mention of a paved inner road which starts at the beginning of the village stretching up to the village Kgotla (*Kgotla* – traditional meeting place). Other positive changes brought about by tourism include solar streets lights and floodlights, which have been erected from tourism revenue. Trees have been planted alongside pavements, and brick and mortar permanent stalls have been built to assist the community during the Khawa tourism annual event or any other activity that would be hosted by the community. There is also a designated camping ground with a gated house. All these are changes in 'villagescape' that have given a new look to the village and enhanced its touristic attractiveness. In addition, the residents indicated that there were some intangible benefits, such as better visibility and awareness of the village:

> The Khawa sand dunes (mounds) as tourist attractions for adventure tourism is now known countrywide, and also are known in neighbouring countries such as in South Africa, Namibia and Zimbabwe, although the event tends to attract more of domestic tourists.

8.4   Conclusions

When CBE is well planned and activities are appropriately managed it has great potential to benefit local people and their associated environment. Several studies have demonstrated the positive effects of community-based ecotourism in the southern Africa region (Makwindi & Ndlovu, 2021; Mbaiwa, 2013; Mearns, 2003; Mmopelwa & Mackenzie, 2020; Monare et al., 2016; Moswete et al., 2012; Saarinen, 2010, 2011; Snyman, 2012). In a similar fashion, the community of Khawa is one such village where residents have formed a CBO/Trust, which runs community-based ecotourism (CBE) projects that aim to benefit community members. In order to boost the tourism impacts, the government introduced an annual tourism event, the Khawa Dune Challenge, which has brought benefits to the people in the form of increased employment. Furthermore, revenue accrued by the community tourism and the Trust is significantly based on tourist camping grounds, tented accommodation, sale of food and horse rides. So far, developments are noticeable as community tourism campsite is upgraded; the main inner road into the village is paved and trees planted for soil conservation. In addition, the residents of Khawa have been empowered through increased participation in and awareness of tourism as a business and they have gained skills and understanding on how to venture into tourism. This indicates that community-based ecotourism can have the potential to unlock socio-economic development challenges in marginalised lands and disadvantaged ethnic communities. These findings are comparable to other studies in Kgalagadi – KD1 Ncaang, Ngwatle and Ukhwi (Arntzen et al., 2003; Moswete et al., 2009; Saarinen et al., 2020), Okavango region (Mbaiwa, 2005, 2008) and North East Botswana (Lenao & Saarinen, 2015). However, like in the other parts of the country, there is an economic leakage problem; the revenue made during the Dune Challenge does not sufficiently trickle down to the community.

Historically, many households were depended on government support in the Khawa area. In this respect, the community and the government have identified tourism as a potential game-changer. However, the study discovered that tourism in the area still operates at a low-key, and it is highly seasonal. Still, the industry has already created some negatively perceived development paths, which need to be proactively managed. A high concentration of tourist activities during the Dune Challenge has resulted in waste disposal problems, which impacts the attractiveness of the village's natural environment – sand dunes, scenic areas, and roadsides – both for tourists and residents. Furthermore, the Khawa village is small, traditional, and surrounded by a fragile desert natural landscape. Large numbers of adventure tourists can have a devastating effect on the environment as it is already being experienced during the Dune Challenge with trampling of the sand mounds by too many people, quad bikers and 4x4 off-road vehicles. Increased tourism has also resulted in noise pollution and social ills. Therefore, more sustainable development-oriented tourism activities and new products are needed within the locality, so that tourism would be more beneficial with inclusive and balanced nature of the tourism development in Khawa.

References

Anderson, W. (2011). Enclave tourism and its socio-economic impact in emerging destinations. *Anatolia, 22*(3), 361–377.

Anup, K. C., Ghimire, S., & Dhakal, A. (2020). Ecotourism and its impact on indigenous people and their local environment: Case of Ghalegaun and Golaghat of Nepal. *GeoJournal*, 1–20. https://doi.org/10.1007/s10708-020-10222-3

Arntzen, J. W., Molokomme, D. L., Terry, E., Moleele, M., Tshosa, T., & Mazambani, D. (2003). *Main findings of the review of community-based natural resources management in Botswana. An occasional paper No. 14*. IUCN/SNV CBNRM support programme.

Borges de Lima, I., & Green, R. (2017). *Wildlife tourism, environmental learning, and ethical encounters: Ecological and conservation aspects*. Springer.

BTDP (Botswana Tourism Development Programme). (2000). *Botswana tourism master plan*. Final Report May 2000). BTDP (Foundation Phase, project N0.7, ACP, BT.4/N0.6 ACP BT 44. Gaborone.

Caldwell, L. (1996). Heritage tourism: A tool for economic development. In P. A. Wells (Ed.), *Keys to the marketplace: Problems and issues in cultural and heritage tourism* (pp. 125–131). Hisarlik Press.

Cater, E. (1993). Ecotourism in the third world: Problems and prospects for sustainable development. *Tourism Management, 14*(2), 85–90.

Central Statistics Office (CSO). (2013). *Botswana population and housing census*. Botswana.

Chiutsi, S., & Saarinen, J. (2017). Local participation in transfrontier tourism: Case of Sengwe community in Great Limpopo Transfrontier Conservation Area, Zimbabwe. *Development Southern Africa, 34*(3), 260–275. https://doi.org/10.1080/0376835X.2016.1259987

Chiutsi, S., & Saarinen, J. (2019). The limits of inclusivity and sustainability in transfrontier peace parks: Case of Sengwe community in Great Limpopo Transfrontier Conservation Area, Zimbabwe. *Critical African Studies, 11*(3), 348–360. https://doi.org/10.1080/2168139 2.2019.1670703

Cobbinah, P., Amenuvor, D., Black, R., & Peprah, C. (2017). Ecotourism in the Kakum Conservation Area, Ghana: Local politics, practice, and outcome. *Journal of Outdoor Recreation and Tourism, 20*, 34–44.

Dhakal, B., & Thapa, B. (2015). Bufferzone management issues in Chitwan national park, Nepal A case study of Kolhuma village development committee. *PARKS, 21*(2), 64–72.

Diallo, I., & Proulx, M. (2016). Socio-economics of aboriginal communities in Quebec. In K. Iankova, A. Hassan, & R. L'Abbe (Eds.), *Indigenous people and economic development: An international perspective* (pp. 223–237). Routledge.

Duim van der, R., Meyer, D., Saarinen, J., & Zellmer, K. (Eds.). (2011). *New alliances for tourism, conservation and development in eastern and southern Africa*. Eburon.

Fennell, D. (2015). *Ecotourism* (4th ed.). Routledge.

Garekae, H., Lepetu, J., & Thakadu, O. T. (2020). Forest resource utilisation and rural livelihoods: Insights from Chobe enclave, Botswana. *South African Geographical Journal, 102*(1), 22–40.

GoB (Government of Botswana). (2003). *National Development Plan 9 2003/04–2008/09*. Ministry of Finance and Development Planning. Gaborone.

GoB (Government of Botswana). (2007). *Community based natural resource management policy. Government paper NO. 2 of 2007*. Government Printer.

Greeffe, X. (2009). Is rural tourism a lever for economic and social development? *Journal of Sustainable Tourism, 2*(1–2), 22–40.

Groves, R. M., Fowler, F. J., Cooper, M. P., Lepkowski, J. M., Singer, E., & Tourangeau, R. (Eds.). (2004). *Survey methodology*. Wiley.

Holliday, A. (2001). *Doing and writing qualitative research*. Sage.

Kavita, E., & Saarinen, J. (2016). Tourism and rural community development in Namibia: Policy issues review. *Fennia, 194*(1), 79–88. https://doi.org/10.11143/4633

Kimaro, M. E., & Saarinen, J. (2020). Tourism and poverty alleviation in the Global South: Emerging corporate social responsibility in the Namibian nature-based tourism industry. In M. T. Stone, M. Lenao, & N. Moswete (Eds.), *Natural resources, tourism and community livelihoods in Southern Africa: Challenges for sustainable development* (pp. 123–142). Routledge.

Koutra, C. (2010). Rapid situation analysis: A hybrid, multi-methods, qualitative, participatory approach to researching tourism development phenomena. *Journal of Sustainable Tourism, 18*(8), 1015–1033.

Lenao, M., & Saarinen, J. (2015). Integrated rural tourism as a tool for community tourism development: Exploring culture and heritage projects in the North-East District of Botswana. *South African Geographical Journal, 97*(2), 203–216.

Lorio, M., & Corsale, A. (2014). Community-based tourism and networking: Viscri, Romania. *Journal of Sustainable Tourism, 22*(2), 234–255.

Makwindi, N., & Ndlovu, J. (2021). Prospects and challenges of community-based tourism as a livelihood diversification strategy at Sehlabathebe National Park in Lesotho. *African Journal of Hospitality, Tourism and Leisure, 10*(1), 333–348. https://doi.org/10.46222/ajhtl.19770720-104

Manning, R. E., Anderson, L. E., & Pettengill, P. R. (2017). *Managing outdoor recreation: Case studies in the national park*. CABI.

Mbaiwa, J. E. (2003). The socio-economic benefits and challenges of a community-based safari hunting tourism in the Okavango Delta, Botswana. *The Journal of Tourism Studies, 15*(2), 37–50.

Mbaiwa, J. E. (2005). Wildlife resource utilisation at Moremi Game Reserve and Khwai community area in the Okavango Delta, Botswana. *Journal of Environmental Management, 77*(2), 144–156.

Mbaiwa, J. E. (2008). *Tourism development, rural livelihoods, and conservation in the Okavango Delta, Botswana, Unpublished PhD dissertation.* Texas A & M University.

Mbaiwa, J. E. (2013). *Community-based natural resource management (CBNRM) in Botswana: CBNRM status report of 2011–2012.* National CBNRM Forum Secretariat and Kalahari Conservation Society.

Mbaiwa, J. E. (2015). Ecotourism in Botswana: 30 years later. *Journal of Ecotourism, 14*(2–3), 204–222.

Mbaiwa, J. E., & Hambira, W. L. (2020). Enclaves and shadow state tourism in the Okavango Delta, Botswana. *South African Geographical Journal, 102*(1), 1–21.

Mearns, K. (2003). Community based tourism: The key to empowering the Sankuyo community in Botswana. *Africa Insight, 33*(1&2), 33–36.

Miles, M. B., Huberman, A. M., & Saldana, J. (1994). *Qualitative data analysis: An expanded sourcebook.* Sage.

Mmopelwa, G., & Mackenzie, L. (2020). Economic assessment of tourism-based livelihoods for sustainable developments: A case of handicrafts in Southern and Eastern Africa. In M. T. Stone, M. Lenao, & N. Moswete (Eds.), *Natural resources, tourism and community livelihoods in southern Africa: Challenges for sustainable development* (pp. 235–253). Routledge.

Monare, M., Moswete, N., Perkins, J., & Saarinen, J. (2016). Emergence of cultural Tourism in Southern Africa: Case studies of two communities in Botswana. In H. Manwa, N. Moswete, & J. Saarinen (Eds.), *Cultural tourism in Southern Africa* (pp. 165–180). Bristol.

Moswete, N. (2009). *Holder perspectives on the potentil for community - based ecotourism and support for the Kgalagadi Transfrontier Park in Botswana.* A PhD dissertation. University of Florida.

Moswete, N., & Mavondo, F. (2003). Problems facing the tourism industry of Botswana. *Botswana Notes and Records, 35,* 69–78.

Moswete, N., Thapa, B., & Lacey, G. (2009). Village-based tourism and community participation: A case study of Matsheng villages in Southwest Botswana. In J. Saarinen, F. Becker, H. Manwa, & D. Wilson (Eds.), *Sustainable tourism in Southern Africa: Local communities and natural resources in transition* (pp. 189–209). Channelview.

Moswete, N., Thapa, B., & Child, B. (2012). Attitudes and opinions of local and national public sector stakeholders towards Kgalagadi Transfrontier Park, Botswana. *International Journal of Sustainable Development and World Ecology, 19*(1), 67–80.

Moswete, N., & Lacey, G. (2014). Women cannot lead: Empowering women through cultural tourism in Botswana. *Journal of Sustainable Tourism, 23*(4), 600–617.

Moswete, N., & Thapa, B. (2015). Factors that influence support for community-based ecotourism in the rural communities adjacent to the Kgalagadi Transfrontier Park, Botswana. *Journal of Ecotourism, 14*(2–3), 243–263.

Moswete, N., & Thapa, B. (2018). Local communities, CBOs/Trusts, and people–park relationships: A case study of the Kgalagadi Transfrontier Park, Botswana. *The George Wright Forum, 35*(1), 96–108.

Moswete, N., Thapa, B., & Darley, W. (2020). Local communities' attitudes and support towards the Kgalagadi Transfrontier Park in Southwestern Botswana. *Sustainability, 12,* 1524.

Mulder, M. B., & Coppolillo, P. (2005). *Conservation: Linking ecology, economics and culture.* Princeton University Press.

Panta, S. K., & Thapa, B. (2017). Entrepreneurship and women's empowerment in gateway communities of Bardia National Park, Nepal. *Journal of Ecotourism, 17*(1), 20–42.

Patton, M. Q. (1999). In 3rde dn (Ed.), *Qualitative research and evaluation methods.* Sage.

Pribudi, A. (2020). Community based approaches to sustainable Batik tourism village area in the special region of Yogyakarta: The case study of Giriloyo village. *Journal of Social Science,* 113–121.

Robins, L. N. (1963). The reluctant respondent. *The Public Opinion Quarterly, 27*(2), 276–286.

Rossman, G. B., & Rallis, S. F. (2017). *An introduction to qualitative research:Learning in the field* (4th ed.). Sage.

Saarinen, J. (2010). Local tourism awareness: Community views on tourism and its impacts in Katutura and King Nehale Conservancy, Namibia. *Development Southern Africa, 27*(5), 713–724.

Saarinen, J. (2011). Tourism development and local communities: The direct benefits of tourism to Ovahimba communities in the Kaokoland, Northwest Namibia. *Tourism Review International, 15*, 149–157.

Saarinen, J. (2017). Enclavic tourism spaces: Territorialization and bordering in tourism destination development and planning. *Tourism Geographies, 19*(3), 425–437.

Saarinen, J., & Manwa, H. (2008). Tourism as a socio-cultural encounter: Host-guest relations in tourism development in Botswana. *Botswana Notes and Records, 39*, 43–53. https://www.jstor.org/stable/41236632

Saarinen, J., Moswete, N., Atlhopheng, J., & Hambira, W. (2020). Changing socio-ecologies of Kalahari: Local perceptions towards environmental change and tourism in Kgalagadi, Botswana. *Development Southern Africa, 37*(5), 855–870.

Saarinen, J., Hambira, W., Manwa, H., & Atlhopheng, J. (2012). Tourism industry reaction to climate change in Kgalagadi South District, Botswana. *Development Southern Africa, 29*, 273–285.

Saarinen, J., & Wall-Reinius, S. (2019). Enclaves in tourism: Producing and governing exclusive spaces for tourism. *Tourism Geographies, 21*(5), 739–748.

Scheyvens, R., & Biddulph, R. (2018). Inclusive tourism development. *Tourism Geographies, 20*(4), 589–609.

Snyman, S. (2012). The role of tourism employment in poverty reduction and community perceptions of conservation and tourism in southern Africa. *Journal of Sustainable Tourism, 20*(3), 395–416.

Southern Africa Development Community Transfrontier Conservation Guidelines: The establishment and Development of TFCA initiatives between SADC member states, www.academia.edu. Retrieved Oct 2020.

Stone, M. T., & Nyaupane, G. P. (2015). Protected areas, tourism and community livelihoods linkages: A comprehensive analysis approach. *Journal of Sustainable Tourism, 24*(5), 673–693.

Stone, M. T., & Stone, L. (2020). Challenges of community-based tourism in Botswana: A review of literature. *Transactions of the Royal Society of South Africa, 75*(2), 181–193.

Strauss, A., & Corbin, J. (1998). *Basic qualitative research: Techniques and procedures for developing grounded theory* (2nd ed.). Sage.

Thusanyo Lefatsheng Trust. (2005). *Land use and management plan: Controlled Hunting Area KD 15. Final Draft.* Khawa Kopanelo Development Trust.

Urquhart, C. (2013). *Grounded theory for qualitative research. A practical guide.* Sage.

Veal, A. J. (2006). *Research methods for leisure and tourism: A practical guide* (3rd ed.). Pearson Educational Limited.

World Bank Group (2021). *Banking on protected areas: Promoting sustainable protected area tourism to benefit local communities.*

Wood, M. E. (2017). *Sustainable tourism on a finite planet: Environmental, business and policy solutions.* Earthscan.

WTO (World Tourism Organisation). (2018). *Tourism and culture synergies.* UNWTO. Retrieved from: https://www.e-unwto.org/. 7 Jan 2021.

Naomi N. Moswete is a Senior Lecturer in the Department of Environmental Science, University of Botswana. Her research interests include Human geography, tourism as a strategy for rural development, community-based tourism; Transboundary conservation areas –ecotourism nexus, parks–people relationships, heritage management & cultural tourism.

Jarkko Saarinen is a Professor of Human Geography (Tourism Studies) at the University of Oulu, Finland, and Distinguished Visiting Professor (Sustainability Management) at the University of Johannesburg, South Africa, and Extraordinary Professor at the Tourism Management Division, Department of Marketing Management, University of Pretoria. His research interests include sustainable development, sustainable tourism, tourism-community relations and nature conservation studies.

Brijesh Thapa is a Professor and Head of the School of Hospitality and Tourism at Oklahoma State University. His research theme is within the nexus of tourism, conservation, and sustainability. He is the Editor-in-Chief for the Journal of Park and Recreation Administration, and a Fellow in The Academy of Leisure Sciences.

Chapter 9
Community-Based Tourism as a Pathway Towards Sustainable Livelihoods and Well-being in Southern Africa

Alinah Kelo Segobye, Maduo Mpolokang, Ngoni Courage Shereni, Stephen Mago, and Malatsi Seleka

9.1 Introduction

Tourism remains a crucial sector in contributing to economic growth and diversification in Africa. In southern Africa, tourism has been critical to economies in the region, especially countries like South Africa and Zimbabwe. As a result, many countries strengthened their tourism policies to become more inclusive and multi-sectoral regarding sustainability and promoting development for communities. These efforts included enhancing heritage sector tourism products and fostering community partnerships in tourism enterprise development. Further, promoting ecotourism and safeguarding natural and cultural heritage became a vital pillar of

A. K. Segobye (✉)
North West University, Mahikeng, South Africa
e-mail: alinah.segobye@gmail.com

M. Mpolokang
Department of Environmental Science, University of Botswana, Gaborone, Botswana

N. C. Shereni
Department of Accounting and Finance, Lupane State University, Lupane, Zimbabwe

School of Tourism and Hospitality, University of Johannesburg, Johannesburg, South Africa

S. Mago
Department of Development Studies, Nelson Mandela University, Gqeberha, South Africa

M. Seleka
Centre for Africa Studies, University of Free State, Bloemfontein, South Africa

© The Author(s), under exclusive license to Springer Nature
Switzerland AG 2022
J. Saarinen et al. (eds.), *Southern African Perspectives on Sustainable Tourism Management*, Geographies of Tourism and Global Change,
https://doi.org/10.1007/978-3-030-99435-8_9

the sector. Community-based tourism (CBT) has become an integral part of both local and national economies through its potential for sustaining the livelihoods of many local communities, that is, supporting small businesses and alleviating poverty. In this regard, tourism serves as an essential contributor to GDP and source of employment for the Southern African Development Community (SADC). Therefore, community-based tourism is promoted as an alternative to private sector-led tourism projects in many parts of the developing world. Botswana's 1990 tourism policy encourages the conservation of natural resources. It focuses on providing local communities with benefits from tourism which encourages communities to appreciate their resources and calls for local communities to share the profits (Government of Botswana, 1990). Similarly, Zimbabwe and South Africa have ensured greater public and private sector participation in the tourism industry through enabling policies and legislation. One critical policy development has been the development of community-based natural resources-based management (CBNRM) as part of broader community-based tourism development strategies.

In this chapter, we highlight the role of community-based tourism in fostering sustainable livelihoods and contributing to development in the SADC region. The SADC region has had the challenge of diversifying the regional economies as countries relied heavily on extractive industries, especially mining. With the declining revenues from mining and global price fluctuations of commodities like copper, it became necessary for the region to explore ways of harnessing the tourism sector to boost revenue and reduce the challenges of unemployment and poverty, especially in rural areas. The chapter explores how initiatives like trans-border frontier parks and CBNRM programmes promote more inclusive development. CBNRM programmes have enabled communities to interact, trade and optimise shared opportunities in the tourism industry, especially in zones of shared cultural and natural landscapes such as the Limpopo and Kasane-Kazungula areas. CBNRM is premised on the principle of decentralisation and democratisation of resources. This includes distributing responsibilities to community members in managing local resources (Sebele, 2010). However, challenges related to community participation include power relations experienced within and between local communities and other stakeholders such as government, businesses, and other actors. The levels and nature of awareness that communities have when undertaking partnerships for CBNRM ventures is essential to reflect on. Evaluations of CBNRM programmes across the SADC region and globally have highlighted how vested and competing interests have marginalised vulnerable groups within communities, including ethnic and indigenous minorities, women, and youth.

We highlight the need to incorporate conflict management and peacebuilding into biodiversity conservation through CBNRM programmes. Due to sometimes conflicting interests of stakeholders, CBNRM initiatives must consider the long-term impacts of the utilisation of resources on communities. In a review of programmes, Botswana identified governance of resources and beneficiation to groups as a potential source of contestation, especially accountability for financial resources accrued in the projects. Efforts ensure that CBT connects Protected Areas (PAs) across international boundaries for ecosystem integrity, functioning, and community identity (Hammill & Besançon, 2003). In turn, considering the complex nexus of PAs and conflict issues, the development of CBT is likely to be contested, affecting sustainable livelihoods within and around the PAs. Therefore, peacebuilding is necessary to ensure interventions can promote local and regional peace and conflict dynamics. Periodic conflicts in the Greater Limpopo and Kasane-Kazungula areas are examples of the necessity of conflict management and peacebuilding in promoting CBT. The chapter is divided into sections, with the first section providing a general background to the topic, briefly introducing the concepts of CBNRM, peacebuilding and sustainable livelihoods. Section two introduces and describes the approach used to gather information. The third section discusses the findings, while the fourth and final section offers concluding remarks.

9.2   CBNRM and Wildlife Management

In southern Africa, the term CBNRM is associated with wildlife and tourism. CBNRM is premised on the principle that if rural communities have the responsibility to manage and benefit from resources in their areas, they are more likely to use them sustainably (Mbaiwa et al., 2019). Therefore, CBNRM initiatives are usually motivated by the central government to local communities with extensive external support. The Brundtland Report of 1987 promoted CBNRM, which emphasised the need to balance the concerns of the poor with conservation imperatives. CBNRM aims to improve natural resource conservation, improve rural livelihoods, reduce poverty and human-wildlife conflicts (Centre for Applied Research, 2016). In the case of Botswana, CBNRM revenues are derived mainly from tourism activities and, to a lesser extent, from sales of veld products. Many Community-Based Organisations (CBOs) have been financially supported at one point in time by government or international cooperating partners (ICPs). In the case of wildlife-based tourism, communities are allocated land for use (e.g. wildlife management areas and/or community land-use zones in protected areas), becoming the source of revenue (Mbaiwa, 2015, 2018; Sebele, 2010).

In the implementation and operation of CBNRM, there is a need to cater to rural development, including devolving decision-making powers from central government to local government and CBOs. There is a need for ongoing policy review and reform to avoid policy contradictions and enable CBOs flexibility to invest in capital ventures at various levels (Dikobe, 2012). What has emerged in four decades of implementing CBNRM programmes suggests competing interests remain and sometimes conflict regarding resources management and fostering partnerships at the community, government and private sector interests in the tourism sector. Greater transparency in managing resources such as licensing wildlife hunting quotas is critical to fostering trust within and between state and citizenry.

9.3 Peacebuilding

Peacebuilding between government and local communities cannot be overempha-
sised, especially in democratic societies. The term 'peacebuilding' was coined in
1992 after a presentation by the late United Nations Secretary-General Boutros-
Ghali. He defined peacebuilding as a range of activities to identify and support
structures that will strengthen and solidify peace to avoid a relapse into conflict
(Boutros-Ghali, 1995). Ultimately, peacebuilding seeks to enhance and promote
human security with the help of democratic governance, human rights, the rule of
law, sustainable development, especially on economic and environmental stewardship and protection to ensure equitable access to resources (Karbo, 2008). African
societies have indigenous resources and institutions for conflict resolution and
peacebuilding. Thus, although there have been challenges as community peacebuilding processes can be time-consuming, peacebuilding initiatives have contributed to community projects.

A literature review on peacebuilding in Africa reveals a limited analytical lens
restricted to post-conflict phases of armed conflict (Ali & Mathews, 2004; Henk,
2005). Peacebuilding including gender, must receive greater attention in discourses
on development in the region given the decades of violent liberation struggles and
post-liberation conflicts. The chapter emphasises the need for post-conflict peacebuilding and peaceful coexistence for communities still experiencing significant
divisions based on race, ethnicity, gender, and class. There is great potential for
sustainable peacebuilding in Africa rooted in indigenous and traditional conflict
resolution mechanisms (Hendricks, 2011; Olonisakin, 2018).

9.4 Tourism and Sustainable Livelihoods

The WCED (1987) defines sustainable development as development that aims to
meet the needs of the present without compromising the ability of future generations to meet their own needs. Sustainable development is premised on economic,
social, and environmental considerations when framing the development agenda. In
this regard, the understanding of sustainability traces back to the Brundtland report
of 1987. Globally, the tourism sector has developed practical ways of assisting all
forms of tourism to move towards sustainability. Saarinen (2006) notes that "sustainability should primarily be connected with the needs of people—not a certain
industry—and the use of natural and cultural resources in a way that will also safeguard human needs in the future (p.1132)". Therefore, community attachment is
important, especially for measuring local support for sustainable tourism development and livelihoods; locals with a bold and robust attachment to place have a
greater concern towards sustainability (Lee, 2013).

Sustainable livelihoods are usually created by identifying avenues in the community which ultimately can generate more sustainable ways for survival. However,

tourism's role in sustainable development highlights the inherent contradictions and complexities of translating notions of sustainability into post-carbon political realities and the centrality of climate change as a sustainable development issue. There is a need for sustainable adaptive measures and contributions to social justice and environmental integrity (Eriksen et al., 2011). We are mindful of the contradictions the notion of sustainable livelihoods elicits when juxtaposed with competitive tourism industries and the needs of communities in PAs or similarly vulnerable and or fragile environments. Again, the concept of community is used fully recognising its ambiguity in the context of people who often have differentiated and competing claims regarding power, resources, and opportunity. We hope to explore some of these intertwined themes through the chapter.

9.5   Methods and Approach

The study adopted a multiple case study inquiry to gain an in-depth understanding of people, environment, and development, focusing on trans-border national parks. Two case studies of communities close to national parks in Zimbabwe and Botswana were purposively selected. These study sites were communities adjacent to Hwange National Park in Zimbabwe and Mababe village between the Moremi Game Reserve and Chobe National Park in Botswana. Primary data was collected from community members, conservation authorities and district administration officials through in-depth interviews (19) and focus group discussions (six) in Mababe village. Literature on the people-parks relationship in peer-reviewed journals, books, government publications, and other relevant materials was also used as data sources to get insight into communities near Hwange National Park.

Further, the chapter draws on broader literature regarding the Great Limpopo Transfrontier Park, which connects South Africa, Botswana, Zimbabwe, and Mozambique (Chiutsi & Saarinen, 2019). The chapter will highlight the intersections between livelihoods and resources as essential ingredients to discourses of peace and security across the SADC region. Securing livelihoods is critical to peace and stability. Securing livelihoods mitigates challenges such as illegal migration across borders and deescalates conflicts where they are emerging and threatening to destabilise tourism and other sectors as indicated by the Botswana -Namibia clashes in the Sedudu/Kasilili area and the Cabo Delgado Province of Mozambique.

The case study approach is appropriate as it allows for a detailed contextual analysis of the people-parks-conflict nexus in communities staying closer to protected areas. In this respect, a comparison of the two case studies enabled us to understand the relationship between communities and protected areas, the nature of conflicts that exist, and the extent to which peacebuilding measures applied to end conflicts have been successful. We used interpretive analysis to identify drivers of conflicts and peacebuilding initiatives in communities bordering protected areas. We documented the lived experience of communities using qualitative methods and, where available, quantitative data sets to corroborate the field interviews data. The following section presents findings from the two case studies, followed by a discussion of the key themes that emerged from them.

9.6 Findings: Case Studies

9.6.1 Case Study: Mababe, Ngamiland District (Botswana)

The village of Mababe is in Ngamiland, northwest Botswana, between Moremi Game Reserve in the south and Chobe National Park in the north. Tourism activities in the area concentrate on wildlife and wilderness experiences with large land areas reserved for wildlife conservation. It offers a case study on the intersections between CBT, sustainable livelihoods, peace and conflict management.

Tourism is a thriving industry due to high touristic interest in Moremi National Park and Chobe National Park, and the Mababe Depression located on the fringes of the village. Approximately twelve (12) tourist vehicles pass through the village and engage in photographic tourism and wildlife viewing daily. Despite this, there are limited benefits to the community, which can be attributed to the foreign ownership of tourism facilities, leading to the export of tourism revenue and the domination of management positions in the hospitality and safari sector by expatriates. Locals get lower salaries with limited opportunities to improve community livelihoods. Protected areas utilised for tourism activities led to the reservation of large chunks of land for the tourism industry's biodiversity conservation and ecosystem services. Tourism-related activities have taken away 'natural capital', which forms the backbone of most community livelihoods in the area.

Restrictive environment conservation policies do not allow communities to gather veld products within protected areas, compromising food security and deepens poverty in the area. The community competes for resources with the tourism industry and, as custodians, do not benefit from the industry, leading to stand-offs between the community and the Department of Wildlife and National Parks (DWNP), which manages the protected areas. Conflict over the alienation of resources from the community in favour of tourism ventures persist.

9.6.1.1 Community-Based Tourism, Conflict and Peacebuilding Efforts in Mababe

Community-based tourism through CBNRM programmes is implemented in Mababe to reduce conflicts and promote peace between the community and DWNP. The programme aims to empower communities and facilitate equitable sharing of ecosystem services benefits. Community-Based Natural Resource Management is a flagship programme that diversifies away from overdependence on traditional land-use patterns to sustainable resource utilisation through tourism. Instead of finding livelihood limitations in conservation activities such as protected areas, the programme has encouraged the community of Mababe to find opportunities in these activities and device ways of establishing livelihoods amid such alienation. The CBNRM in Mababe is implemented through Mababe Zokotsama Community Development Trust (MZCDT). The Trust was awarded controlled hunting area NG41 by DWNP and engages in tourism projects based on their natural resources. The Trust has sought business partners to set up lodges and camps in NG41, creating employment for the community. Locals acknowledge that through CBNRM programmes, their lives have slightly improved, and they also have a sense of ownership of tourism activities. It was also evident that the community makes decisions regarding what to do and with whom to partner. Furthermore, CBNRM facilitates peace because the locals have supplemented their traditional livelihood activities and rely on tourism ventures, thus reducing conflicts over resources.

However, it is essential to note that though CBNRM has fostered peace in Mababe, there are disaffected community members. Some community members felt that CBNRM programmes follow fragmented implementation and management processes. They cited interference from DWNP, particularly in decision-making processes. Another issue raised was that the community has not been capacitated with entrepreneurship and management skills, making it challenging to fully harness the programme's economic benefits. Governance of the community Trust raised concerns of corruption. The community respondents stated that projects were awarded to close associates who do not fulfil their obligations and instead keep profits. Others cited examples of the Trust owing monies and benefits to former employees despite the availability of financial resources.

The community acknowledged the potential of CBNRM in fostering peace between them and DWNP. They cited capacity building and knowledge resources as needed to effectively manage projects to enhance benefits, increase commitment to projects, and secure their participation in promoting the sector and sustaining the environment.

9.6.2 Case Study: Hwange National Park (Zimbabwe)

9.6.2.1 Collaborative Resource Management in Communities Adjacent to Hwange National Park, Zimbabwe

Hwange National Park (HNP) is located in Matabeleland North Province in Hwange district and is the largest Protected Area in Zimbabwe, covering an area of about 14,650 km^2 (Muboko et al., 2014). The Park, currently managed by the Zimbabwe Parks and Wildlife Management Authority, was designated a game reserve in 1928 (Guerbois et al., 2013), then known as Wankie Game Reserve. The Protected Area (PA) has grown to be a significant resource in wildlife-based tourism in Zimbabwe and falls in the Victoria Falls Tourism Development Zone because of its closeness and importance to the tourism resort town (MoTHI, 2016). The National Park shares a fenceless border with Botswana allowing for free movement of animals between HNP and wildlife sanctuaries on the Botswana side. HNP is part of the Kavango–Zambezi Transfrontier Conservation Area (KAZA), which includes other protected areas from Zambia, Botswana, Namibia and Angola. Hwange district comprises 20 wards that are near the national park. The ethnic composition of people in Hwange includes Nambiya, Tonga, Ndebele, Shona, Nyanja, Dombe, Lozvi and others (Shereni & Saarinen, 2020).

9.6.2.2 Human-Wildlife Conflict in Hwange

Humans and animals share water and grazing land, which frequently leads to human-wildlife conflict. Increasing human population and expanding settlements into wildlife corridors have increased human-wildlife conflict in Hwange District (Guerbois et al., 2013). The scholars also noted that even though extraction of resources like thatching grass is allowed under close monitoring by park authorities, contentions arise from restricted access to natural resources in HNP. In addition, changes in land-use patterns due to the land reform programme resulting in people occupying areas previously reserved for wildlife exacerbate conflict between humans and wildlife. The CAMPFIRE Association (2020) observed that the movement of wildlife from HNP into community areas presents challenges to community members, such as destruction of crops, depredation of livestock, spread of diseases, and loss of human life. Without meaningful benefits cascading to the locals accruing from the wildlife resources, locals see wild animals as a menace.

   In most cases, community members react by poaching the wild animals or facilitating poaching by syndicates involved in illegal hunting to compensate for their losses. There have been reports of numerous poaching incidents in communities adjacent to Hwange national park. The most talked-about was cyanide poisoning in 2013, which killed more than 300 elephants and many more animal and bird species (Muboko et al., 2014). Human-wildlife conflict increases when the burden exerted by animals exceeds the benefits reaped by the community, which is currently the case in communities adjacent to HNP.

CAMPFIRE in Hwange district Communal Area Management Programme for Indigenous Resources (CAMPFIRE) is a CBNRM initiative introduced in Zimbabwe in the late 1980s. It is of the earliest CBNRM practices in Southern Africa (Roe & Nelson, 2009). The programme has been implemented in various rural communities around Zimbabwe to benefit community members settled close to wildlife conservation areas (Dube, 2019). CAMPFIRE aimed to provide a mechanism where local community members can manage and benefit from resources in their localities to start seeing wild animals as an asset rather than a menace. In the past, CAMPFIRE in Hwange has seen benefits such as the construction of social infrastructure, funding of income-generating projects and access to game meat accruing to the community members (Sce Shereni & Saarinen, 2020). Eighteen (18) out of 20 wards in

Hwange District are recognised as CAMPFIRE wards (CAMPFIRE Association, 2020), and three wards, Mabale (17), Sidinda (8) and Silewu (15), are the wildlife producer wards where most of the CAMPFIRE activities take place (Dube, 2019). The Hwange Rural District Council (RDC), which links local communities and central government, manages CAMPFIRE in Hwange. An elected village and ward CAMPFIRE committee, which makes decisions on behalf of the community members (Balint & Mashinya, 2006), represents local communities.

Income generated from the CAMPFIRE programme in Hwange is derived mainly from trophy hunting fees, CBT projects, and fees from leasing properties. According to CAMPFIRE Association (2020), revenue sharing guidelines follows that 55% of the income goes to communities and 26% to RDCs to support costs associated with running the CAMPFIRE programme. The RDCs retain 15% for administration 4% as a levy to the association. Muzirambi et al. (2020) noted that the RDCs do not adhere to the revenue sharing guidelines in some instances. They tend to retain most of the revenue and pass a small amount to the communities mainly because the policies are not legally binding. The distribution of income to community members is to atone for the destruction caused by wild animals and incentivise communities to conserve wildlife. Table 9.1 shows the income distribution from CAMPFIRE in Hwange District between 2009 and 2016.

Currently, CAMPFIRE in Hwange District faces numerous challenges such as lack of community participation in decision making, failure to devolve authority to local communities, inequitable distribution of benefits and corruption by the CAMPFIRE committee, among others (Shereni & Saarinen, 2020). Challenges faced in CAMPFIRE communities have seen benefits accruing to locals diminishing greatly to a point where the community members do not see the need for such a programme in their area (Muzirambi et al., 2020). The decline in tourist arrivals and Zimbabwe's economic crisis have compounded the challenges facing CAMPFIRE programmes, further reducing benefits to local communities. The case study notes that the local communities can get tremendous benefits from the resources within

Table 9.1 Distribution of income from CAMPFIRE

Year	Gross income (USD)	Community 55% (USD)	CAMPFIRE management 26% (USD)	Council levy 15% (USD)	CAMPFIRE association 4% (USD)
2009	32,500	17,874	8450	4875	1300
2010	41,725	22,948	10,848	6258	1669
2011	63,070	34,648	16,398	9460	2522
2012	74,408	40,924	19,346	11,161	2976
2013	65,300	35,915	16,978	9795	2612
2014	85,777	47,177	22,302	12,866	3431
2015	49,350	27,142	12,831	7402	1974
2016	31,450	12,978	7293	10,057	1122

Source: CAMPFIRE Association, 2020

the Park, which can help incentivise conservation efforts and reduce people-Park management conflicts.

9.7 Discussion

The case studies above reveal that communities settled close to wildlife areas expect to improve their livelihoods from the natural resources within their proximity. The notable boom of tourism activities in protected areas is seen as how local communities can benefit through employment creation, the establishment of community-based tourism enterprises, and supporting CBNRM initiatives (Lenao & Saarinen, 2015). The case studies show that communities are not benefitting as much as they should from the resources in their areas because of several factors that include corruption, lack of community involvement in tourism activities and benefits not accruing directly to households (Shereni & Saarinen, 2020). It is also clear that communities near protected areas bear the brunt of human-wildlife conflict because wildlife strays into their settlements (Amaja et al., 2016). Numerous challenges facing communities near protected areas include livestock depredation, destruction of crops, loss of human lives and spread of diseases from wildlife (Schnegg & Kiaka, 2018). The low level of benefits to locals, the challenges they face in their communities, and the menace brought about by wildlife lead to conflicts and tensions in wildlife areas. CBNRM practices, when seen as conflict resolution mechanisms and promoted, can ensure that communities benefit from wildlife resources and help defuse tensions in communities (Mbaiwa et al., 2019). The CAMPFIRE programme and community trusts in the case study areas can provide a lot of benefits to the locals if they are managed well (DeGeorges & Reilly, 2009). This study, therefore, noted that the involvement of rural communities in managing resources in their areas through initiatives such as community-based tourism helps achieve sustainable livelihoods.

9.8 Conclusion and Recommendations

This chapter highlighted the interconnections between CBT, sustainable livelihoods, conflict, and peace management in southern Africa. The nexus between community livelihoods and tourism emerges from the case studies. Interventions through policies and programming can mitigate poverty and other challenges or exacerbate these. The chapter noted that sustainability depends on many factors that influence the performance of CBT. The current economic challenges and security concerns that impact tourism have further reduced opportunities for communities dependent on CBT. From the case studies, there are similarities in community responses and government interventions. Employment beneficiation for the local communities has proven to be a common feature in CBT. Both cases have shown that conflict management and peacebuilding at local levels are central to CBT. Perceptions of good resources governance are also crucial to promoting peace. In the interest of regional development, developing administrative and legal frameworks to address resources sharing, prevent conflicts, and foster inter-country resource management remains a challenge, especially with increased security threats from new conflicts driving people displacement and insecurity. Despite these challenges, CBT, if implemented well, facilitates community-managed tourism. The use of natural resources for tourism will enable communities to use resources sustainably and add value to the tourism sector through the diversification of tourism and economic opportunities (Rozemeijer et al., 2012).

The case studies provide insights into CBT's benefits, especially for rural communities in developing countries. Despite the challenges facing communities living in PAs, such as threats from increased poaching activity and reduced access to resources, CBT remains an attractive means of sustaining livelihoods and promoting inclusive development. Human-wildlife conflict, environmental changes and governance which affect sustainable resources management can be mediated through peacebuilding especially managing micro conflicts before they escalate. Skills development in creative industries for youth and women can augment community skills. Policy and legal reforms in and between states are essential for promoting people and wildlife's movement and security. We see the Africa Continental Free Trade Agreement's enactment and implementation as a milestone towards fostering regional cooperation likely to benefit tourism and sustainability (Kende-Robb, 2021). Therefore, future strategies and plans for CBT will define whether it remains a distinctive and unique form of tourism in southern Africa that can contribute to local communities' cultural and heritage management and indigenous knowledge.

The chapter advances the following recommendations, which factor in the dynamic nature of the sector considering the adverse impacts of the COVID-19 pandemic on the industry, communities, and livelihoods. CBNRM and CBT are critical to the sustainable management of natural and cultural heritage in southern Africa. It is crucial to provide targeted support for community-based organisations as they are vital stakeholders in CBT. Non-state actors such as CSR/CSI initiatives and development partners should regularly assess the impact of programmes. Addressing governance issues in state-led institutions and CBOs is critical to ensuring conflict can be prevented and managed. Promoting peace between stakeholders and the local citizenry can be enhanced through democratising resources management and inclusive governance. Policy and legislative review are critical to ensure equity and access to foster an enabling environment. More research is needed in the region to encourage the involvement of CBT in heritage management to enhance the sustainable use of natural and cultural heritage resources. Ultimately, the governments in the region must foster regional cooperation, and international protocols for sustainable development should be domesticated to ensure local protection of people and resources.

References

Ali, T., & Mathews, R. (2004). *Durable peace: Challenges to peacebuilding in Africa*. University of Toronto Press.

Amaja, L. G., Feyssa, D. H., & Gutema, T. M. (2016). Assessment of types of damage and causes of human-wildlife conflict in Gera district, south western Ethiopia. *Journal of Ecology and the Natural Environment, 8*(5), 49–54. https://doi.org/10.5897/JENE2015.0543

Balint, P. J., & Mashinya, J. (2006). The decline of a model community-based conservation project: Governance, capacity, and devolution in Mahenye, Zimbabwe. *Geoforum, 37*(5), 805–815. https://doi.org/10.1016/j.geoforum.2005.01.011

Botswana Government. (1990). Botswana Tourism Policy. Gaborone, Government Printers.

Boutros-Ghali, B. (1995). *Agenda for peace* (2nd ed.). United Nations.

CAMPFIRE Association Zimbabwe. (2020, June 2020). *Community benefits summary*. Available at: https://www.campfirezimbabwe.org/article/community-benefits-summary

Centre for Applied Research. (2016). *Review of community based natural resources management in Botswana*. Centre for Applied Research.

Chiutsi, S., & Saarinen, J. (2019). The limits of inclusivity and sustainability in transfrontier peace parks: Case of Sengwe community in Great Limpopo transfrontier conservation area, Zimbabwe. *Critical African Studies, 11*(3), 348–360.

DeGeorges, P. A., & Reilly, B. K. (2009). The realities of community-based natural resource management and biodiversity conservation in Sub-Saharan Africa. *Sustainability, 1*, 734–788. https://doi.org/10.3390/su1030734

Dikobe, L. (Ed.) (2012). *Natural resources at the centre of rural livelihoods –looking beyond 50 years of Botswana's independence*. Proceedings of the 7th biennial national CBNRM conference. Gaborone: Botswana CBNRM National Forum.

Dube, N. (2019). Voices from the village on trophy hunting in Hwange district, Zimbabwe. *Ecological Economics, 159*, 335–343. https://doi.org/10.1016/j.ecolecon.2019.02.006

Eriksen, S., Aldunce, P., Bahinipati, C. S., Martins, R. D., Molefe, J. I., Nhemachena, C., & Ulsrud, K. (2011). When not every response to climate change is a good one: Identifying principles for sustainable adaptation. *Climate and Development, 3*, 7–20.

Guerbois, C., Dufour, A.-B., Mtare, G., & Fritz, H. (2013). Insights for integrated conservation from attitudes of people toward protected areas near Hwange national park, Zimbabwe. *Conservation Biology, 27*(4), 844–855. https://doi.org/10.1111/cobi.12108

Hammill, A., & Besançon, C. (2003). Promoting conflict sensitivity in transboundary protected areas: A role for peace and conflict impact assessments. In *Transboundary protected areas in the governance stream of the 5th world parks congress* (pp. 12–13). Durban.

Hendricks, C. (2011). *Gender and security in Africa: An overview*. Nordic Africa Institute.

Henk, D. (2005). The Botswana defence force and the war against poachers in Southern Africa. *Small Wars and Insurgencies*, 170–191.

Karbo, T. (2008). Peace-building in Africa. In D. J. Francis (Ed.), *Peace and conflict in Africa* (pp. 113–147). Zed Books.

Kende-Robb, C. (2021). *6 reasons why Africa's new free trade area is a global game-changer*. https://www.weforum.org/agenda/2021/02/afcfta-africa-free-trade-global-game-changer/

Lee, T. H. (2013). Influence analysis of community resident support for sustainable tourism development. *Tourism Management, 34*, 37–46. https://doi.org/10.1016/j.tourman.2012.03.007

Lenao, M., & Saarinen, J. (2015). Integrated rural tourism as a tool for community tourism development: Exploring culture and heritage projects in the North-East District of Botswana. *South African Geographical Journal, 97*(2), 203–216. https://doi.org/10.1080/03736245.2015.1028985

Mbaiwa, J. E. (2015). Community-based natural resource management in Botswana. In R. van der Duim, M. Lamers, & J. van Wijk (Eds.), *Institutional arrangements for conservation, development and tourism in Eastern and Southern Africa* (pp. 59–80). Springer.

Mbaiwa, J. E. (2018). Effects of the safari hunting tourism ban on rural livelihoods and wildlife conservation in northern Botswana. *South African Geographical Journal, 100*(1), 41–61.

Mbaiwa, J. E., Mbaiwa, T., & Siphambe, G. (2019). The community-based natural resource management programme in Southern Africa-promise or peril? In M. Mkono (Ed.), *Positive tourism in Africa* (pp. 11–22). Routledge.

MoTHI. (2016). *National tourism master plan.* https://drive.google.com/file/d/1BJT76E8iojCJ9t9 fM78W1pcVXIpxv0Ox/view.

Muboko, N., Muposhi, V., Tarakini, T., Gandiwa, E., Vengesayi, S., & Makuwe, E. (2014). Cyanide poisoning and African elephant mortality in Hwange National Park, Zimbabwe: A preliminary assessment. *Pachyderm, 55*(55), 92–94.

Muzirambi, J. M., Musavengane, R., & Mearns, K. (2020). Revisiting devolution in community-based natural resources management in Zimbabwe: Towards inclusive governance approaches. In M. T. Stone, M. Lenao, & N. Moswete (Eds.), *Natural resources, tourism and community livelihoods in southern Africa* (pp. 143–158). Routledge.

Olonisakin, F. (2018). Towards re-conceptualising leadership for sustainable peace. *Leadership and Developing Societies, 2*(1), 1–30.

Roe, D., & Nelson, F. (2009). The origins and evolution of community-based natural resource management in Africa. In D. Roe, F. Nelson, & C. Sandbrook (Eds.), *Community management of natural resources in Africa: Impacts, experiences and future directions* (pp. 5–12). International Institute for Environment and Development.

Rozemeijer, N., Gujadhur, T., Motshubi, C., Van den Berg, E., & Flyman, M. V. (2012). *Community based tourism in Botswana.* Gaborone.

Saarinen J. (2006). *Traditions of sustainability in tourism studies annals of tourism research, 33*(4), 1121–1140.

Schnegg, M., & Kiaka, R. D. (2018). Subsidised elephants: Community-based resource governance and environmental (in)justice in Namibia. *Geoforum, 93,* 105–115. https://doi.org/10.1016/j. geoforum.2018.05.010

Sebele, L. S. (2010). Community-based tourism ventures, benefits and challenges: Khama rhino sanctuary trust, Central District, Botswana. *Tourism Management, 31*(1), 136–146.

Shereni, N. C., & Saarinen, J. (2020). Community perceptions on the benefits and challenges of community-based natural resources management in Zimbabwe. *Development Southern Africa.* https://doi.org/10.1080/0376835X.2020.1796599

WCED. (1987). *Our common future.* Oxford University Press.

Alinah Kelo Segobye is the Dean of Faculty – Human Sciences at the Namibia University of Science and Technology (NUST). She is an Extraordinary Professor at North West University in the Indigenous Knowledge Centre. She is an affiliated Research Scholar at the African Futures Institute (AFI) in Pretoria, South Africa. Segobye has teaching, research and consultancy experience in the areas of African studies, including culture; heritage management, HIV/AIDS; gender and development. Segobye has authored and co-authored several essays and book chapters on Africa's development.

Maduo Mpolokang holds a MSc in Environmental Science from the University of Botswana with a keen interest in tourism development, climate change and environmental sustainability management. He is currently a Trainee Environmental Assessment Practitioner under Aqualogic Pty (Ltd), Botswana.

Ngoni Courage Shereni is a faculty member at Lupane State University in Zimbabwe. Currently, he is pursuing a PhD in Tourism and Hospitality at the University of Johannesburg in South Africa. His research interests are in sustainable tourism, Sustainable Development Goals (SDGs) in the Tourism and Hospitality industry, tourism and hospitality education, disruptive technology in the tourism industry, tourism exhibitions as well as Community-based Natural Resources Management (CBNRM) practices, among others.

Stephen Mago is an Associate Professor of Development Studies and Head of Department at Nelson Mandela University, South Africa. He chairs the Faculty Research Ethics Committee at NMU. He is a multidisciplinary researcher who likes collaborating with researchers from different fields of study. His specific research interests are development finance, entrepreneurship, research methodology, local economic development and rural development. He serves as Guest Editor for a Special Issue for the Management and Economics Research Journal (MERJ) titled "Global Economic Issues".

Malatsi Seleka is a Doctoral Candidate (Africa Studies: Peace and Conflict in Context) at the Centre for Africa Studies, University of Free State, South Africa. He holds a Master's Degree in Development Studies and a BA Degree in Archaeology from the University of Botswana. Currently, he works as an External Community Development Consultant at the Institute of Development Management, Gaborone. His research interests include natural resources management, heritage preservation, sustainable development practices, indigenous knowledge systems, land management, climate change adaptation, governance and conflict management.

Chapter 10
Changing Environment and the Political Ecology of Authenticity in Heritage Tourism: A Case of the Ovahimba and the Ju/'Hoansi-San Living Museums in Namibia

Isobel Green and Jarkko Saarinen

10.1 Introduction

Namibia is the driest country in southern Africa. The country has been facing recurrent drought since 2013, characterised by below-average or no rainfall at all and soaring temperatures. The year 2019 was reported as the driest in 90 years of recorded history. Due to this situation, the government of Namibia has declared three times a state of emergency between 2013 and 2019 (Keja-Kaereho & Tjizu, 2019). Droughts have been highly problematic, especially in the rural parts of the country, which are highly dependent on livelihoods based on the direct use of natural resources and the conditions of ecosystem services (Bollig, 2016).

Many ethnic groups have traditionally adapted to the changing environmental conditions, but because of the intensified droughts, they face increasing challenges to practice their traditional livelihoods (see Keja-Kaereho & Tjizu, 2019; Saarinen, 2016). Recently, a key alternative livelihood option for rural communities in Namibia has been the evolving tourism industry, which is increasingly used as a tool for local development and economic diversification in the country (Kalvelage et al., 2020). As a result, tourism has been introduced to local communities (Lapeyre,

I. Green (✉)
Department of Tourism and Hospitality, Namibia University of Science and Technology, Windhoek, Namibia

University of Johannesburg, School of Tourism and Hospitality, Johannesburg, South Africa
e-mail: igreen@nust.na

J. Saarinen
Geography Research Unit, University of Oulu, Oulu, Finland

School of Tourism and Hospitality, University of Johannesburg, Johannesburg, South Africa
e-mail: jarkko.saarinen@oulu.fi

© The Author(s), under exclusive license to Springer Nature   139
Switzerland AG 2022
J. Saarinen et al. (eds.), *Southern African Perspectives on Sustainable Tourism Management*, Geographies of Tourism and Global Change,
https://doi.org/10.1007/978-3-030-99435-8_10

2010, 2011a; Novelli & Gebhardt, 2007), and the government has a proactive approach with regards to policies to integrate communities in tourism development and related societal modernisation process (Kavita & Saarinen, 2015; Lapeyre, 2011b).

One specific tool for integrating local communities into tourism, especially ethnic minorities living in remote areas, has been the establishment of so-called living museums in Namibia. Living museums are an exhibitionary museological format, which relies on the stimulation of everyday practices and is characterised by live interpretation and performances (Naumova, 2015). They have their antecedents in the open-air museums in northern Europe and North America (see Corsane, 2004). Living museums offer a tactile experience of culture and history while creating both narrative and physical space for visitors to interact with the cultural production of heritage. Thus, compared to many other forms of museums that deal with physical objects, these living museums are more concerned with safeguarding everyday intangible cultural heritage. This everyday heritage refers to the social practices of a culture and, thus, not necessarily material objects characterising a specific culture and ethnic group.

According to Naumova (2015) living museums are often referred to as heritage museums. However, a heritage museum can be considered as a more generic form that encompasses any kind of museum that is dedicated to the preservation of a local past significant to the surrounding community and culture. Thus, a living museum is a specific form of museology consisting of heritage with spaces that look as though they are inhabited (Naumova, 2015). Gordon (2016) has defined a living museum as a cultural institution that teach historical lessons by recreating past environments that explicitly use interpreters to demonstrate past ways of life. This may lead to a danger that living museums are considered as unchanged portraits of "an idealised past" (Romo, 2010, p. 10). However, Walter (2020) has further argued that living museums could also be tourism-related community-based projects designed, run, and staffed by community hosts. They could also be interactive and changing community-based exhibitions controlled by the community. While living museums are produced and designed spaces for touristic experiences, the idea of authenticity forms a crucial dimension for them (Williams, 2013). In this respect, living museums represent how people have lived and continue to live with their environment.

In addition to living museums, there are some other forms of heritage tourism sites based on a local community dimension in southern Africa. Cultural villages, for example, are heritage tourism attractions manifesting regional policy programs aiming to use local culture both in economic growth and community empowerment (van Veuren, 2001). Compared to living museums, cultural villages are an older form of heritage tourism attractions in the region. The majority of cultural villages in southern Africa, especially in South Africa, Namibia, Botswana and Swaziland, were established in the 1990's (Saarinen, 2007) with a strong emphasis "on the potential of cultural tourism to contribute towards the goals of sustainable rural development" (van Veuren, 2004, p. 139). They are purposefully built attractions that do not necessarily aim to demonstrate an authentic, i.e. real and living local village, but a staged place for cultural performances, information sharing and learning.

According to Chhabra et al. (2003, p. 702) "much of today's heritage tourism product depends on the staging or re-creation of ethnic or cultural traditions". This may problematise the connections between authenticity and living museums as heritage tourism attractions (Lenao & Saarinen, 2015; Timothy, 2014). This paper discusses how the heritage elements are produced and displayed at living museums by utilising an example of the Ovahimba and the Ju/'Hoansi-San living museums in Namibia. Both ethnic groups are visibly used in Namibian tourism marketing and form a core attraction for international tourists (Saarinen & Niskala, 2009). Furthermore, the chapter analyses how the displayed heritage tourism and its authenticity have been affected by environmental changes at the case study sites. The chapter utilises a political ecology perspective to understand the intertwined nature of local culture, ways of living and the environment (Bryant & Bailey, 1997). According to Robbins (2012) the political ecology lens can be used to understand the decisions that local communities make about the natural environment. Specifically, political ecology is used in this paper to discuss how environmental change has impacted the ways communities' position and depict themselves and work with tourism. This paper gives an overview of the ideas and connections of heritage tourism and authenticity, followed by the case study examples and a concluding section. The results are based on fieldwork undertaken at the Ovahimba and Ju/'Hoansi-San Living Museums between 2017 and 2019, utilising an observation approach in the local communities at the living museums. The description of the study sites is based on the fieldwork notes and observations.

10.2 Heritage, Tourism, and Authenticity

10.2.1 Heritage Tourism

Heritage tourism concerns the motivation to experience traditions, customs and artefacts representative of past and present time at a tourist destination (Park et al., 2019). Though there are different views on what constitutes heritage, a common aspect is that it is something that is traditionally associated with what is inherited, valued, wanted and handed from one generation to the next (Herbert, 1995; Hoyau, 1988; Walsh, 1992). Graham (2002) states that heritage is the contemporary use of the past, including its interpretations and its representations. As stated by Park (2013), heritage is neither fixed nor unchanging but socially produced, conditioned, and constantly negotiated.

Heritage can be both tangible and intangible. In contrast to tangible material traces of past, intangible heritage refers to the practices, representations, expressions, and the knowledge and skills that communities, groups and in some cases individuals, recognise as part of their cultural heritage. According to the 2003

Convention for the Safeguarding of the Intangible Cultural Heritage (UNESCO, 2003), the intangible cultural heritage (ICH) can be defined as traditional and contemporary living at the same time, mainly recreated orally, and the skills often shared and performed by communities. However, the distinction between tangible and intangible heritage is not always distinctive. The interpretation of heritage can involve both tangible and intangible heritage elements by using the original objects, skills and enacting the practices to communicate information of the respective culture to the tourist.

Tourism has been a vital tool for utilising heritage resources commercially (Timothy, 2014). Many tangible heritage sites have become popular tourist attractions (Lak et al., 2020; Smith, 2016), while intangible heritage has been more challenging to commercialise in tourism. Heritage tourism can be defined as visitation to a historical area consisting of activities that provide a historical experience with educational value based around consumer motivation (Carter & Horneman, 2001; Goh, 2010; Prentice, 1993; Zeppel & Hall, 1992). There are two schools of thought in heritage tourism research, either emphasising the role of heritage resources or the role of tourist experiences. On the one hand, Fyall and Garrod (1998), for example, define heritage tourism as an economic activity that makes use of socio-cultural assets to attract visitors. This refers to the place-based idea of heritage tourism.

On the other hand, Poria et al. (2001, p. 1047) see heritage tourism as an activity based on visitors' motivations and perceptions rather than on specific places and site attributes (see Brida et al., 2012; Li et al., 2016). Similarly, Zeppel and Hall (1992) highlight the role of tourist motivation as a search for experiences based on nostalgia. These latter views are more focused on the motivational or customer-driven basis of heritage tourism.

Heritage tourism can involve both dimensions, i.e. consumption of experiences and creating site-specific facilities, products, and resources. There is also a need to develop tourism services and infrastructure beyond the site scale in the form of roads and transportation, for example (Brooks, 2011). From a geographical perspective, heritage tourism is based on the heritage values of a place or region and is inherently place-specific and stems from the unique character of the place (Tuan, 1974). Thus, the identity of the place is marketed with integration to the tourism products that enable tourists to experience and appreciate the place and its attributed. In tourism, a key element for this process of 'appreciation' is heritage interpretation, which Tilden (1977, p. 8) has defined as an "educational activity which aims to reveal meanings and relationships through the use of original objects, by firsthand experience and by illustrative media rather than simply to communicate factual media" (Smith, 2016; Uzzell, 1989). Schouten (1995) has further argued that visitors at heritage sites are looking for experiences rather than the hard facts of historical reality. This may problematise the connection between heritage tourism products and experiences and their authenticity.

10.2.2 Authenticity in Tourism

Authenticity is one of the most challenging concepts and ideas in tourism studies (Cohen, 1988; MacCannell, 1976), and there is no clear consensus regarding the definition of the term (Li et al., 2016). Originally, the term authenticity was mainly linked to museums and their artefacts (Trilling, 1972). In tourism, the term usually refers to a wider set of issues and processes that are beyond the objective criteria of museums (Wang, 1999). Provocatively stated, there may be nothing real or authentic in tourism (Boorstin, 1964) as a large share of the products and performances in tourism are what Brown (1996, p. 33) calls "genuine fakes". Obviously, this depends on what we mean by authenticity and whether we consider that authentic forms of culture and traditions can change.

In this respect, Wang (1999) has approached the authenticity in tourist experiences as three different types: objective, constructive and existential authenticity. According to him, objective authenticity is based on modernism and interprets authenticity as the objective character of sites and traditions. It refers to the museum-linked usage of authenticity with an absolute and objective criterion used to measure authenticity. Constructive authenticity is based on constructivism. It argues that authenticity is a result of social constructs based on worldviews, beliefs, knowledge, perceptions, etc., that are value-based, transforming and influenced by power relations (Cheong & Miller, 2000; Saarinen, 1999; Urry, 1990). Constructed authenticity is often based on stereotyped images and expectations onto toured objects (Wang, 1999), which MacCannell (1973) has called staged authenticity. Finally, existential authenticity refers to highly subjective views towards tourist sites and performances (Williams, 2013). As noted by Wang (1999, p. 359): "existential authenticity can often have nothing to do with the issue of whether toured objects are real."

Previous research has shown that the elements linked to authenticity may change over time (Blapp & Mitas, 2019; Saarinen, 2007). Cultures are dynamic (Burns, 2001), and they have always been in contact with and influenced by other cultures and places. Thus, a general modernisation process of societies, environmental change, globalisation, and tourism development, for example, all create potential changes in our everyday living and traditions (Azarya, 2004). Growing heritage tourism can have significant influences on how the everyday life of locals is structured, performed, evolves and adapts to other changes (Qu et al., 2019). On the one hand, it can help to preserve cultural elements as they have 'always been', which is implicitly the idea and objective of living museums. On the other hand, it can accelerate changes and modernisation as a high number of tourists visiting heritage sites may change local ways of life (Blapp & Mitas, 2019). Therefore, instead of focusing on a fixed idea of authenticity in tourism, some scholars have suggested moving beyond authenticity discourse in tourism studies and focusing on "the local way of life" (Blapp & Mitas, 2019, p. 36) that is changeable and includes the elements of the past, the present and potential futures (Saarinen, 2007).

10.3   The Ovahimbas and the Ju/'Hoansi San

The Ovahimba live in the north-western part of Namibia, in the Kunene region, and southwestern Angola (Bollig & Heinemann, 2002; Saarinen, 2012, 2013). Traditionally, the Ovahimba have been pastoral nomads whose main livelihoods are based on cattle and goats and seasonal small-scale subsistence crop farming. Women tend to perform labour-intensive work next to villages, such as carrying water, building homes, and milking cows. Ovahimba men handle the livestock, political tasks, and legal trials in traditional law-making. Their houses are relatively simple, cone-shaped structures of saplings, bound together with palm leaves, mud and cow dung that offer shade and coolness in a dry and warm environment. The women rub their bodies with *otjize*, a red mixture of butterfat and ochre, believed to protect their skins against the sun. The red mixture is said to symbolise earth's rich red colour and the blood that symbolises life. The Ovahimba still adorn themselves with traditional jewellery according to traditional customs. Both men and women wear large numbers of necklaces and arm bracelets made from ostrich eggshell beads, grass, cloth, and copper (Shilongo, 2020).

The Ju/'Hoansi San have been researched by anthropologists for over 50 years (Barnard, 2007; Biesele, 1986; Gordon & Douglas, 2000; Lee, 1986; Marshall, 1976) and are probably among the most studied and best documented indigenous peoples in Africa. Currently, around 30,000 to 33,000 San live in Namibia. Traditionally, the San people have lived a hunting and gathering economic lifestyle and are nomadic, i.e. highly mobile, and they relocate themselves in response to their seasonal subsistence needs. As a result, their nomadic movements have been driven by environmental conditions, and they have been dependent on wild plants for their day-to-day dietary needs. The Ju/'Hoansi belong to the San (formerly known as the Bushmen), a collective name for the 14 Khoesan-speaking groups whose languages are characterised by its clicking sounds. The Ju/'Hoansi are the second biggest San group in Namibia situated around the area of Tsumkwe.

Due to intensified droughts in Namibia, many local communities have had to look to other means to diversify their traditional livelihood as the natural environment could no longer meet the needs of their nomadic lifestyles (Saarinen, 2016). The community's livestock has been diminishing ever since and has reached a point where they can no longer keep large herds of livestock. The wild fruits are not as abundant as before, which has affected the traditional lifestyles of many Ovahimba and San communities. With this dwindling of natural resources and reduced capacity for dependence on the ecosystem, these two case study communities, jointly with the Living Culture Foundation (2020), have set up the Living Museums to aid community members to supplement their subsistence farming. As a result, communities have transformed by moving away from living solely based on their natural resources.

10.4   Case Study Sites and Research Materials

The Ovahimba Living Museum and the Ju/'Hoansi-San Living Museum are designed as places where tourists can interactively experience the traditional culture of the Ovahimba and the San. These living museums act as 'traditional schools for guests' but also for the younger generations of the communities. At the same time, the museums are income-generating institutions for the communities (Living Culture Foundation, 2020). The museums offer performances, other activities and dedicated guides who interpret and explain the various activities that take place at the sites.

Over the years, the Ovahimba communities have been exposed to both tourists and the practices of the tourism industry in the region (Saarinen, 2012, 2016). The Kunene region, where the Ovahimba mainly are, has several conservancies (e.g. Marienfluss, Orupembe, Sanitatas, Okondjombo, Puros and Kunene River), and this has resulted in the increase in tourism activities in the region (Shilongo, 2020), including the Ovahimba Living Museum where 71 visitors in 2016 and 512 in 2017 visited the Ovahimba Living Museum. (Living Culture Foundation, 2020). The museum, established in 2016, is strategically located between Opuwo, the administrative capital of Kunene Region, and the Epupa Falls, one of the key tourist attractions in northwest Namibia. There is an entrance fee of between N\$ 120 and N\$ 4000 per person, depending on the program visitors select. The Museum area consists of a large traditional homestead in which the Ovahimbas introduce interested guests to their daily routines and encourage them to participate in various activities. Some of the performed traditions are the production of food, craftsmanship (forging, pottery, wood carving, leather tanning), the building of clay huts, and singing and dancing (Living Culture Foundation, 2020).

The Ju/'Hoansi-San Living Museum, referred to as /Xao-o Ju/'Hoansi-Ga (i.e. the life of the Ju/'Hoansi), is situated in Grashoek, a small village, about halfway between Grootfontein and Tsumkwe, north of the C44 road. Like the Ovahimba Living Museum, the Ju/'Hoansi-San Living Museum is an open-air museum where the community demonstrates their lifestyle and interacts with tourists by showcasing their various daily activities. Originally initiated by the Namibian tour guide Werner Pfeifer and the teacher GhauNaici from Grashoek (Living Culture Foundation, 2020) it has been operated by the community since July 2004. Its annual visitor numbers fluctuate, but in 2017 there were 2899 visitors in the Museum (Living Culture Foundation, 2020).

Many San groups still practice the hunter-gatherer lifestyle. The museum aims to provide visitors with insights into the life of the San by presenting an old but slowly disappearing culture in their reconstructed 'nomad-village'. The activities at the constructed Ju/'Hoansi-San Living Museum include gathering food from the bush and collecting various plant species to make traditional medicines. They also use ostrich shells to make the various traditional crafts and prepare animal skins to make their clothing and leather bags that they sell at the museum site. Almost all offered programs are interactive. Visitors try to shoot arrows, experience the unique

Ju/'Hoansi rope skipping or try to sing an original song. By doing all this, the museum aims to set a high value on presenting the hunter-gatherer culture as authentically as possible. The Living Museum of the Ju/'Hoansi-San makes use guides to translate for visitors from the native San Language to English during their visit to the museum.

This research is grounded on a qualitative approach with an observation method based on interactions with the local community members at the Living Museums. The fieldwork took place at the Ovahimba and Ju/'Hoansi-San Living Museums from 2017 till 2019. There were approximately 20 local community members at both research sites. Various data collection methods were utilised but the fieldwork materials used in this chapter are solely based on unstructured observations (Bell & Bryman, 2011). The fieldwork observations included interactions between local community members and visitors at the Living Museums. The analysis is based on a general narrative and ethnographically informed approach where the focus is exploring the life of the local community members and how the performed heritage elements may have changed due to droughts in the study areas.

10.5 Heritage Tourism and Authenticity in the Ovahimba and the Ju/'Hoansi-San Living Museums

10.5.1 Visiting Living Museums

The daily routine of the local communities, as observed at the Ovahimba Living Museum and the Ju/'Hoansi San Living Museum, are the main activities promoted for potential visitors. These activities include making traditional crafts and ornaments, gathering plants for making traditional medicine and making traditional weapons such as bow and arrows, above the normal duties of cleaning the homestead and preparing meals.

In addition, especially in the Ovahimba Living Museum, the observed activities included women doing each other's or their own hair and making the traditional perfume that they wear. They also made clothes from animal skin. The Ovahimba's key routine activity is keeping the place clean, where they show the visitors their cultural performances and rituals. The Ovahimba ladies also prepare traditional medicines from plants they collect nearby for visitors to observe.

Traditionally, the Ovahimba have relied on their livestock and gardening maze and/or vegetables. During the observation period, however, there was very little livestock around, and women were no longer working in the fields. During harsh times like the persistent drought at the site, livestock needs to be constantly located in new areas with better grazing and water. Thus, environmental conditions directly affect their livelihood resources and how they can demonstrate their traditions and way of life to visitors. Due to the drought, the local community members mainly performed activities that were not directly related to the natural environment but instead showed visitors craft making and explaining about the Ovahimba rituals, for example.

During the observations and discussions, the local communities indicated that once tourists come to the Living Museum, the local communities give the tourists a program of activities that they can participate in to choose from. This applied to both Living Museums. Then the hosts would prepare activities that the tourists would like to engage in. Once the tourists decide which activities they want, the local community would recreate this event or ritual for them. This 'order model' for cultural rituals performed created a constructed authenticity. As the local people had no capacity to practice their everyday routines with livestock and garden due to the drought, the living museum ideal turned towards a cultural village concept with purely performed and staged authenticity.

In the Ju/'Hoansi San Living Museum, the daily activities undertaken by the local community included collecting various plant species to make traditional medicines and collecting different berries and plants for food. They make use of animal skins in preparing their clothing and leather bags. The bags are sold at the Living Museum in a designated place for souvenirs, normally a few meters outside the area where the community members would reenact their cultural activities. Locals also use ostrich shells to make various traditional crafts that they also sell to visitors or wear themselves when the next visitors come to the Living Museum.

Based on the observations at the Ju/'Hoansi Living Museum, a similarly constructed authenticity was performed but in a slightly differently staged site. When entering the place, a visitor needs to go to the information meeting point where the locals wait for them. Based on the visitors' preferences of activities, the locals select who will accompany the visitor(s) to the Living Museum, which is about 0.5 km from the meeting point. This contrasts with the Ovahimba Living Museum, where the locals live and perform the activities on site. This separation of living elsewhere and performing the activities at the Ju/'Hoansi resembles a typical staged cultural village.

10.5.2 Drought and Performed Authenticity of Heritage Tourism

During the persistent drought of 2017–2019, the Namibian government declared a state of emergency. Due to the drought, many communities had to adapt their way of life, including the communities around the Ovahimba and the Ju/'Hoansi San Living Museums. In 2017, before the intensified drought in the Ovahimba Living Museum, the men and herd boys worked with the livestock, and women did gardening and ploughing in the fields. This was done whether tourists were there or not. In 2019, the men no longer tended livestock, and women did not work in the fields. The drought had influenced the traditional way of life. Thus, due to environmental change, they adapted and altered their behaviour. This has impacted the nature of authenticity, as the living museum ideal with normal everyday living and routines has transformed towards forms typical for cultural villages in which staged performances take place and life is a tourism centric in general.

While the Ju/'Hoansi San Living Museum has been more staged designed, as people do not actually live at the site, tourists visit and performances take place, also there the drought period and related environmental changes have had impacted the way of life of the local community. The local community has not been able to showcase some of the traditional activities, such as taking visitors to show various plant species as they do not exist in the surroundings due to droughts. This environmental change, however, has provided an opportunity for the local communities to strengthen their commercial heritage tourism activities, although in a staged manner. In many respects, these cultural heritage interactions are no longer being done to preserve their culture but to supplement the daily life of locals influenced by negative environmental changes. Therefore, heritage tourism and related guiding have become a 'job' and potential career path for the local community members in which practices their children and younger generations are actively engaged.

10.6   Conclusions

Drought and tourism development have brought change to local ways of living in Namibia and the case study sites. The findings here indicate that local communities have transformed their lifestyles towards a more 'modern' way of life that involves the active production and consumption of heritage tourism experiences within a monetary economy. The communities have become more tourism-dependent as they can no longer live a traditional way of life that has been traditionally based on natural resources alone. They have learnt to supplement their lifestyle by engaging in tourism activities. What kind of identity issues this change may cause in future, however, is unknown. As the children are intensively socialised into heritage tourism and related experiences, the political ecology of droughts and changed authenticity may transform the community identities more dramatically in future than the environmental change alone would do. As noted by Robbins (2012, p. 23), environmental change and conditions can "lead to new kinds of people" with opportunities and constraints.

In respect to a heritage tourism experience, this involves potentially problematic issues. If visiting tourists expect objectively authentic experiences from the Living Museums based on nostalgia, as it is ideally indicated in the place promotion, it may have become an unrealistic target and basis for marketing in future. The Living Museums are increasingly providing staged, i.e. constructed and performed authenticity, which is the typical form of authenticity in the production circles of the tourism industry in general (Williams, 2013). As a result, the Living Museums have become cultural villages designed to satisfy the cultural and heritage consumption needs of the visitors and be a key source of income and employment for the locals. Still, these sites can also be interpreted as representing the current 'real local of a way of life' under the changing environment and societies in the Global South (Walter, 2020). They are not static but dynamic. In this respect, the Living Museums may still work for the local cultures to preserve their traditions while transforming and adapting to the 'new normal'. Positively, this can be considered as being resilient.

Droughts have affected the natural environment and the traditional way of life in the case study areas. The communities have diversified to supplement their traditional lifestyles with tourism practices and use tourism for community development based on their own needs with performed authenticity. From a community perspective, this is neither positive nor negative change, per se. All cultures are dynamic and cannot remain stagnant. Thus, these communities must adapt to the changing environment and their surrounding society. At the same time, however, we need to believe, trust and support that the Ovahimba and the Ju/'Hoansi San communities in Namibia can use tourism in a considerate manner to provide socio-economic benefits, empowerment and protect the cultures in the ways they prefer. By doing so, the tourism industry could better serve the current calls for sustainable development.

References

Azarya, V. (2004). Globalisation and international tourism in developing countries: Marginality as a commercial commodity. *Current Sociology, 52*(6), 949–967. https://doi.org/10.1177/0011392104046617

Barnard, A. (2007). *Anthropology and the bushman*. Berg Publishers.

Bell, E., & Bryman, A. (2011). *Research methodology: Business and management contexts*. Oxford University Press.

Biesele, M. (1986). How hunter-gatherers' stories make sense: Semantics and adaption. *Cultural Anthropology, 1*(2), 157–170.

Blapp, M., & Mitas, O. (2019). The role of authenticity in rural creative tourism. In N. Duxbury & G. Richards (Eds.), *A research agenda for creative tourism* (pp. 28–41). Edward Elgar Publishing.

Bollig, M. (2016). Towards an arid Eden? Boundary-making, governance and benefit sharing and the political ecology of the new commons of Kunene region, Northern Namibia. *International Journal of the Commons, 10*(2), 771–799. https://doi.org/10.18352/ijc.702

Bollig, M., & Heinemann, H. (2002). Nomadic savages, ochre people and heroic herders: Visual presentation of the Himba of Namibia's Kaokoland. *Visual Anthropology, 15*, 267–312.

Boorstin, D. (1964). *The image: A guide to pseudo-events in America*. Vintage Book.

Brida, J., Disegna, M., & Scuderi, R. (2012). The visitor's perception of authenticity at the museums: Archaeology versus modern art. *Current Issues in Tourism, 17*(6), 518–538.

Brooks, G. (2011). *Heritage as a driver for development*. ICOMOS. Available at: http://openarchive.icomos.org/id/eprint/1207/1/III-1-Article1_Brooks.pdf

Brown, D. (1996). Genuine Fakes. In Selwyn, T. (Eds.), *The Tourist Image, Chichester* (pp. 33–47). John Wiley and Sons.

Bryant, R., & Bailey, S. (1997). *Third world political ecology*. Routledge.

Burns, P. (2001). Brief encounters: Culture, tourism, and the local-global nexus. In S. Wahab & C. Cooper (Eds.), *Tourism in the age of globalisation* (pp. 290–305). Routledge.

Carter, R., & Horneman, L. (2001). Does a market for heritage tourism exist? *Journal of the Australasian Institute for Maritime Archaeology, 25*, 61–68.

Cheong, S.-M., & Miller, M. L. (2000). Power and tourism: A Foucauldian observation. *Annals of Tourism Research, 27*(2), 371–390.

Chhabra, D., Healy, R., & Sills, E. (2003). Staged authenticity and heritage tourism. *Annals of Tourism Research, 30*(3), 702–719.

Cohen, E. (1988). Authenticity and commoditisation in tourism. *Annals of Tourism Research, 15*(3), 371–386.

Fyall, A., & Garrod, B. (1998). Heritage tourism: At what price? *Managing Leisure, 3*(4), 213–228. https://doi.org/10.1080/136067198375996

Goh, E. (2010). Understanding the heritage tourist market segment. *International Journal of Leisure and Tourism Marketing, 1*(3), 257–270.

Gordon, A. (2016). *Time travel: Tourism and the rise of the living history museum in mid-twentieth century*. University of British Columbia Press.

Gordon, R., & Douglas, S. (2000). *The bushmen myth: The making of a Namibian underclass*. Westview Press.

Corsane, G. (Ed.). (2004). *Heritage, museums and galleries*. Routledge.

Graham, B. (2002). Heritage as knowledge: Capital or culture? *Urban Studies, 39*(5–6), 1003–1017. https://doi.org/10.1080/00420980220128426

Herbert, G. (1995). *The critical heritage*. Routledge.

Hoyau, P. (1988). Heritage and the conserver society: The French case. In R. Lumley (Ed.), *The museum time-machine* (pp. 27–35). Comedia/Routledge.

Kalvelage, L., Revilla Diez, J. R., & Bollig, M. (2020). How much remains? Local value capture from tourism in Zambezi, Namibia. *Tourism Geographies*. https://doi.org/10.1080/1461668 8.2020.1786154

Kavita, E & Saarinen, J. (2015). Tourism and rural community development in Namibia: Policy issues review. *Fennia, 193*(3), 1–10.

Keja-Kaereho, C., & Tjizu, B. (2019). Climate change and global warming in Namibia: Environmental disasters vs human life and the economy. *Management and Economics Research Journal, 5*(1), 1–11.

Lak, A., Gheitasi, M., & Timothy, D. J. (2020). Urban regeneration through heritage tourism: Cultural policies and strategic management. *Journal of Tourism and Cultural Change, 18*(4), 386–403. https://doi.org/10.1080/14766825.2019.1668002

Lapeyre, R. (2010). Community-based tourism as a sustainable solution to maximise impacts locally? The Tsiseb conservancy case, Namibia. *Development Southern Africa, 27*, 757–772.

Lapeyre, R. (2011a). The Grootberg lodge partnership in Namibia: Towards poverty alleviation and empowerment for long-term sustainability? *Current Issues in Tourism, 14*, 221–234.

Lapeyre, R. (2011b). Governance structures and the distribution of tourism income in Namibia communal lands: A new institutional framework. *Tijdschrift voor Economische en Sociale Geografie, 102*, 302–315.

Lee, R. (1986). The gods must be crazy, but the state has a plan: Government policies towards the San in Namibia. *Canadian Journal of African Studies, 20*(1), 91–98.

Lenao, M., & Saarinen, J. (2015). Integrated rural tourism as a tool for community tourism development: Exploring culture and heritage projects in the North East District of Botswana. *South African Geographical Journal, 97*(2), 203–216. https://doi.org/10.1080/0373624 5.2015.1028985

Li, X., Shen, H., & Wen, H. (2016). A study on tourists perceived authenticity towards experience quality and behaviour intention of cultural heritage in Macao. *International Journal of Marketing Studies, 8*(4), 117–123.

Living Culture Foundation. (2020, August 1). Traditional cultures in Namibia. Retrieved from: https://www.lcfn.info/

MacCannell, D. (1973) Staged Authenticity: Arrangement of Social Space in Tourist Settings. *American Journal of Sociology, 79*(3).

MacCannell, D. (1976). *The tourist*. Schocken Books.

Marshall, L. (1976). *The Kung of Nyae Nyae*. Harvard University Press.

Naumova, A. (2015). Touching the past: Investigating lived experiences of heritage in living history museums. *The International Journal of the Inclusive Museum, 7*(3–4), 1–9.

Novelli, M., & Gebhardt, K. (2007). Community based tourism in Namibia: 'Reality Show' or 'Window Dressing'? *Current Issues in Tourism, 10*, 443–479.

Park, H. (2013). *Heritage tourism*. Routledge.

Park, E., Choi, B., & Lee, T. (2019). The role and dimensions of authenticity in heritage tourism. *Tourism Management, 74*(2), 99–109.

Poria, Y., Airey, D., & Butler, R. (2001). Challenging the present approach to heritage tourism: Is tourism to heritage places heritage tourism? *Tourism Review, 56*(1–2), 51–53. https://doi.org/10.1108/eb058358

Qu, C., Timothy, D. J., & Zhang, C. (2019). Does tourism erode or prosper culture? Evidence from the Tibetan ethnic area of Sichuan Province, China. *Journal of Tourism and Cultural Change, 17*(4), 526–543. https://doi.org/10.1080/14766825.2019.1600867

Prentice, R. (1993). *Tourism and heritage attractions*. Routledge.

Robbins, P. (2012). *Political ecology: A critical introduction*. Wiley-Blackwell.

Romo, A. A. (2010). *Brazil's living museum: Race, reform, and tradition in Bahia*. North Carolina Press.

Saarinen, J. (1999). Representations of indigeneity: Sami culture in the discourses of tourism. In P. M. Sant & J. N. Brown (Eds.), *Indigeneity: Constructions and re/presentations*. Nova Science Publishers.

Saarinen, J. (2007). Cultural tourism, local communities and representations of authenticity: The case of Lesedi and Swazi cultural villages in Southern Africa. In B. Wishitemi, A. Spenceley, & H. Wels (Eds.), *Culture and community: Tourism studies in eastern and southern Africa* (pp. 140–154). Rozenberg.

Saarinen, J. (2012). Tourism development and local communities: The direct benefits of tourism to OvaHimba communities in the Kaokoland, North-West Namibia. *Tourism Review International, 15*, 149–157.

Saarinen, J. (2013). Ethnic tourism in Kaokoland, northwest Namibia. In G. Visser & S. Ferreira (Eds.), *Tourism and crisis* (pp. 180–194). Routledge.

Saarinen, J. (2016). Political ecologies and economies of tourism development in Kaokoland, North-West Namibia. In M. Mostafanezhad, A. Carr, & R. Norum (Eds.), *Political ecology of tourism: Communities, power and the environment* (pp. 213–230). Routledge.

Saarinen, J., & Niskala, M. (2009). Local culture and regional development: The role of OvaHimba in Namibian tourism. In P. Hottola (Ed.), *Tourism strategies and local responses in southern Africa* (pp. 61–72). CABI Publishing.

Schouten, F. (1995). Improving visitor care in heritage attractions. *Tourism Management, 16*(4), 259–261.

Shilongo, A. (2020). Tourism and commoditisation of traditional cultures among the Himba people of Namibia. *Editon Consortium Journal of Arts, Humanities, and Social Studies, 2*(1), 187–197.

Smith, M. (2016). *Issues in cultural tourism studies*. Routledge.

Tilden, J. (1977). *Interpreting our heritage: Principles and practices for visitors in parks, museums, and historic places*. North Carolina Press.

Timothy, D. (2014). Contemporary cultural heritage and tourism: Development issues and emerging trends. *Journal of Public Archaeology, 13*(1–3), 30–47.

Trilling, L. (1972). *Sincerity and authenticity*. Butterworth-Heinemann.

Tuan, Y. (1974). *Topophilia: A study of environmental perception, attitudes and values*. Columbia University Press.

UNESCO (United Nations Educational, Scientific and Cultural Organization). (2003). *Convention for the safeguarding of the intangible cultural heritage*. UNESCO.

Urry, J. (1990). *The tourist gaze: Leisure and travel in contemporary societies*. Sage.

Uzzell, D. (1989). Introduction: The visitor experience. In D. Uzzel (Ed.), *Heritage interpretation* (pp. 1–15). Belhaven.

van Veuren, E. J. (2001). Transforming cultural villages in the spatial development initiatives of South Africa. *South African Geographical Journal, 83*(2), 137–148.

van Veuren, E. J. (2004). Cultural village tourism in South Africa: Capitalising on indigenous culture. In C. M. Rogerson & G. Visser (Eds.), *Tourism and development issues in contemporary South Africa*. Africa Institute of South Africa.

Walsh, K. (1992). *The representation of the past: Museums and heritage in the postmodern world.* Routledge.

Walter, P. (2020). Community-based ecotourism projects as living museums. *Journal of Ecotourism, 19*(3), 233–247.

Wang, N. (1999). Rethinking authenticity in tourism experience. *Annals of Tourism Research, 26*(2), 349–370.

Williams, P. (2013). Performing interpretation. *Scandinavian Journal of Hospitality and Tourism, 13*(2), 115–126.

Zeppel, H., & Hall, C. (1992). Arts and heritage tourism. In B. Weiler & C. Hall (Eds.), *Special interest tourism* (pp. 47–68). Belhaven Press.

Isobel Green is a PhD. Candidate at University of Johannesburg and Master of Leisure Project Management, University of Deusto, Bilbao, Spain. Her research interests are heritage, cultural and creative tourism, women and indigenous people.

Jarkko Saarinen is a Professor of Human Geography (Tourism Studies) at the University of Oulu, Finland, and Distinguished Visiting Professor (Sustainability Management) at the University of Johannesburg, South Africa, and Extraordinary Professor at the Tourism Management Division, Department of Marketing Management, University of Pretoria. His research interests include sustainable development, sustainable tourism, tourism-community relations and nature conservation studies.

Chapter 11
Perspectives on the Applicability of Nexus Thinking to Private Protected Areas: A Case Study of Mokolodi Nature Reserve, Botswana

James Maradza, Raban Chanda, and Naomi N. Moswete

11.1 Introduction

The need for solutions to global environmental challenges, such as climate change, loss of biodiversity, global warming, and poverty, continues to be the priority of research across the globe. However, the advent of Community Based Natural Resource Management (CBNRM) has been welcomed as an innovative way of addressing socio-economic and ecological challenges (GoB, 2007; Mbaiwa, 2011). Indeed, the strategy made remarkable achievements in terms of resource utilisation and management, but its applicability is limited to the management and utilisation of common-pool resources (Blaikie, 2006; DeGeorges & Reilly, 2009; Hoole, 2008; Phuthego & Chanda, 2004; Swatuk, 2005). It is the complexity of appropriating benefit to local communities from privately owned natural resources that demand a holistic, interconnected and multifaceted approach of which CBNRM is devoid of.

Keeping in perspective that the future prospects of protected areas are limited without the involvement and support of the local communities (Beresford & Phillips, 2000), coupled with the inter-linkages existing between protected areas and the local people considering the importance of the linkages to livelihoods, ecotourism and conservation; the proverbial question is how benefits can be channelled to the local communities in a privately owned protected ecotourism conservation area. The chapter aims to study the perception of stakeholders on the challenges and opportunities of the applicability of nexus thinking (NT), within this context, as a

J. Maradza · N. N. Moswete (✉) · R. Chanda
Department of Environmental Science, University of Botswana, Gaborone, Botswana
e-mail: CHANDAR@ub.ac.bw; moatshen@ub.ac.bw

© The Author(s), under exclusive license to Springer Nature
Switzerland AG 2022
J. Saarinen et al. (eds.), *Southern African Perspectives on Sustainable Tourism Management*, Geographies of Tourism and Global Change,
https://doi.org/10.1007/978-3-030-99435-8_11

multifaceted model that offer equitable benefits of conservation from the support and involvement of the local communities, without upsetting the functioning of privately owned and managed resources. The realisation of the connection between biodiversity conservation, tourism and livelihoods (Nexus) brings in the NT model as a collaborative sustainable approach to address the challenges of poverty, loss of biodiversity and environmental degradation. The study argues that the stakeholder's perceptions on the applicability of NT are an important step that will help to establish the opportunities and challenges of the NT model. The study provides an empirically informed position analysis and further enriches the discourse of the applicability of NT on private protected areas gathered from key informant interviews of Mokolodi Nature Reserve's stakeholders.

11.2 Literature Review

11.2.1 Tourism Context

Tourism has become one of the fastest-growing economic industries in the world and the third-largest industry after chemicals and fuels (UNWTO, 2018; Christian et al., 2011). Globally, tourism generates 11% of Global Gross Domestic Product (GDP) and employs over 200 million people, accounting for one in every ten jobs (Crotti & Misrahi, 2017). Internationally, arrivals of tourists increased from 2.5 million in 1950 to 1.186 billion in 2015, with an average record of over 800 million international travellers every year and the growth is expected to surpass 1.8 billion by 2030 (Crotti & Misrahi, 2017). Therefore, if tourism is properly managed, it has the potential to fulfil the three dimensions of sustainable development, namely: social, economic and environmental. At least, Agenda 2030 for Sustainable development Goals (SDGs) – notably Goals 8, 12 and 14 – appreciate tourism's role in fostering economic growth for sustainable livelihoods.

In Botswana, tourism is the second-largest economic sector, after diamonds mining and processing (Mopelwa & Blignaut, 2014), contributing 9.7% towards the country's GDP (WTTC, 2009), creating employment to approximately 13,000 (Rabaloi, 2006). The unique natural resources and the renowned World Heritage Sites like the Okavango Delta protected under the Ramsar Convention – have a catalytic influence for economic and social growth (GoB, 2001; Mopelwa & Blignaut, 2014). Based on these positive attributes, it is imperative to continue soliciting sustainable, innovative ideas that boost socio-economic and ecological development in Botswana, as envisaged in the country's National Development Plan (NDP 11) and Vision 2036.

11.2.2   The Poverty Versus the Environment Debate

Botswana is an upper-middle-income country, with the national average poverty rate at 16.3% in 2015 (Central Statistics Office (CSO), 2018; World Bank, 2020). Unfortunately, poverty is directly linked to environmental degradation in a vicious circle. The absence of alternative means of livelihood, especially in marginalised sections of the society, leaves the poor with no option except to plunder natural resources for survival. Unless and until poverty alleviation strategies provide social and economic benefits to the local people, poverty and environmental degradation remains a challenge, especially in developing countries (Matseketsa et al., 2018; Snyman, 2013). According to Robertson (1989), human poverty is the inability of the people to afford average standards of living: access to food, clean water, shelter. Poverty can be defined either in absolute or relative terms. In relative terms, Laderchi et al. (2003) describe it as a modern Eurocentric construct that clustered countries together as poor on the basis that their overall income is insignificant as compared to those countries dominating the world's economy. In absolute terms, the World Bank defines poverty in terms of one's income level, which is $1.25 per day (Laderchi et al., 2003; World Bank, 2020). Although the concept of poverty alleviation dates back to the 1970s, it was only viewed as a theme of development in the 90s after the World Bank's World Development Report on Economic Structural Adjustment Programme (ESAP) (Culpeper, 2005). ESAP was a Poverty Reduction Strategy (PRS) to ease debts and secure international funding in developing countries (Culpeper, 2005). However, to date, poverty reduction remains a yet to be achieved goal.

Lately, tourism has been in the international spotlight as a possible alternative to poverty alleviation. This led to the poverty reduction initiative: Sustainable Tourism-Elimination of Poverty (ST-EP), launched by The World Tourism Organization in Johannesburg in 2002, to channel small to medium tourism projects to marginalised rural communities (Adams et al., 2004; see Snyman, 2013; Suich et al., 2015). Likewise, Batswana's tourism initiatives connect conservation with livelihoods as a means to poverty alleviation and biodiversity conservation. However, this is only applicable when benefits from the exploitation of resources are made constant with the present needs of the local people (Mbaiwa et al., 2019).

11.2.3   Community-Based Natural Resource Management: An Overview

The strategy of integrating human resources into natural resource conservation is not new in Africa. A recount of colonial history records revealed that legendary communities of the Maasai -Mara, the Ngorongoro and the Amboseli areas in East Africa practised sustainable utilisation of natural resources (Murphree, 1998). Similarly, in pre-colonial Botswana, communities satisfactorily managed their own natural resources according to their traditional *dikgosi* customs, and knowledge (Arntzen et al., 2007; Phuthego & Chanda, 2004). Unfortunately, today, critical observation and literature reveal that most of the sub-Saharan African communities are battling with loss of biodiversity and environmental degradation, heavily impacting the poor (Schlossberg et al., 2019; Stone; 2013). Natural resources are dwindling both in absolute numbers and diversity. Estimates from a global assessment conducted in 2014 show that from the 71,576 terrestrial and freshwater species assessed: 860 were extinct; 21,286 were threatened, and 4286 were critically endangered (Pimm et al., 2014).

Early efforts to address these challenges led national governments in partnership with international organisations like USA Agency for International Development (USAID) to prioritise local communities' participation in rural conservation development projects (Suich et al., 2015). This marked the beginning of CBNRM as a new paradigm of resource utilisation and conservation philosophy. The concept of CBNRM was not only credited for its commitment to ensuring that local people are involved in the management and conservation of natural resources (Mbaiwa et al., 2019), but it prided itself as a crucial livelihood benefactor in rural development (Moswete & Thapa, 2018; Sebele, 2010). CBNRM can be equated to a people-centred participatory resource management strategy with the prime focus of achieving sustainable and equitable use of local natural resources (Arntzen et al., 2007). The approach gained remarkable supremacy after the publication of the World Commission on Environment and Development (Brundtland) Report in 1987. Regionally, CBNRM was first established in Zimbabwe through (CAMPFIRE) in the 1980s. By late 1980, the approach had spread across the whole of Southern Africa (Mbaiwa et al., 2019; Nyaupane & Poudel, 2011).

Botswana hailed CBNRM as a panacea to biodiversity depletion after suffering, for so many decades, from significant declines in biodiversity (Mbaiwa et al., 2011; Sebele, 2010). According to Steiner and Rihoy (1995), the CBNRM approach was meant to counteract issues of species extinction due to population growth, overuse of resources and human-wildlife conflict. Communities were motivated to actively participate in the management of natural resources through Community Based Organisations (CBOs) known as Trusts and Community Based Tourism (CBT). The success of CBNRM was noted in Zimbabwe's CAMPFIRE. In their remarks, Mutandwa, and Gadzirayi (2007) concluded that CAMPFIRE empowered rural communities with the "rights to manage", "rights to benefit", and "rights of disposal of natural resource utilisation".

However, noble as it is, studies widely acknowledged that CBNRM was limited to the governance of common-pool resources (Blaikie, 2006; Fabricius & Collins, 2007; Phuthego & Chanda, 2004). Mbaiwa (2011) concurs that the paradigm of CBNRM was built upon common property theory, that common-pool resources can be sustainably utilised when community autonomy is recognised. In that respect, CBNRM is not applicable to both PPAs and GPAs such as national parks, except in buffer zones known as Wildlife Management Areas (WMAs) in Botswana (Armitage, 2005). This current discourse treats CBNRM as a less robust institutional arrangement that, at best, can only buffer the protection of common property resources (Hoole, 2008). This shortcoming of CBNRM requires further research to solicit collaborative sustainable approaches as models that are best applicable to PPAs and GPAs. The nexus thinking (NT) approach has opted as a novel framework that could add insight and knowledge in promoting community support for conservation and sustainable utilisation of natural resources. This is the much-anticipated scientific research framework capable of addressing the contemporary limitation of CBNRM. Since stakeholders are important components in any given institution, it was obligatory to establish the stakeholder's perception about the novel NT framework.

11.2.4   The Concept and Relevance of Nexus Thinking

The term nexus refers to a bond or link connecting members of the group (Groenfeldt, 2010). In support Leck et al. (2015) define nexus as one or more connections linking two or more things. Both definitions capture the interactive and linkage notion effect that characterises cooperation, coordination, interdependence for long term development. It provides a practical platform for novel strategies that portray cross-sectorial, multi-scale interdependencies that reduce mismatches in decision making, planning and management, thereby increasing synergies and promoting resource security (Bizikova et al., 2013; WEF, 2012). It is, therefore, a strategic and holistic style of thinking that considers long-term implications across interlinked areas, weighing up and balancing socio-economic and environmental goals.

The linkages avert possibilities of instability or crisis arising because of mismanagement, overexploitation or unsustainable utilisation of resources. It was only in 2008, during the World Economic Forum (WEF), that the concept of NT was revealed as a response to the call for action on water resource management (WEF, 2012). During the drafting of SDGs by the United Nations, it is worth noting that the NT concept was their point of reference (Hussey & Pittock, 2012). There is, therefore, no doubt that the incorporation of NT in global strategic working plans like SDGs validates the relevance and necessity of utilising the concept in any development strategy. The greatest novelty of the NT concept is reflected in the ability to connect socio-economic development aspects with natural ecosystems protection as a win-win solution (Benson et al., 2015). However, NT has limitations: it fails to deal with shifts in relation to global markets and natural phenomena such as the current Covid-19 global pandemic, recession, and climate change (Scoones, 2009). Furthermore, the proposed stakeholder dialogue may fail to engage the poorest or most vulnerable members of society. However, the study submits that these limitations are sufficiently addressed by NT framework as it is a holistic approach.

The NT conceptual framework (Fig. 11.1) is formulated from the literature of previous studies reviewed on the interactions between a Private Protected Area (PPA) and local communities and how such interactions shape sustainable livelihoods, ecotourism and conservation outcomes. It consists of four main components: Communal Area Communities, Ecotourism, Conservation and

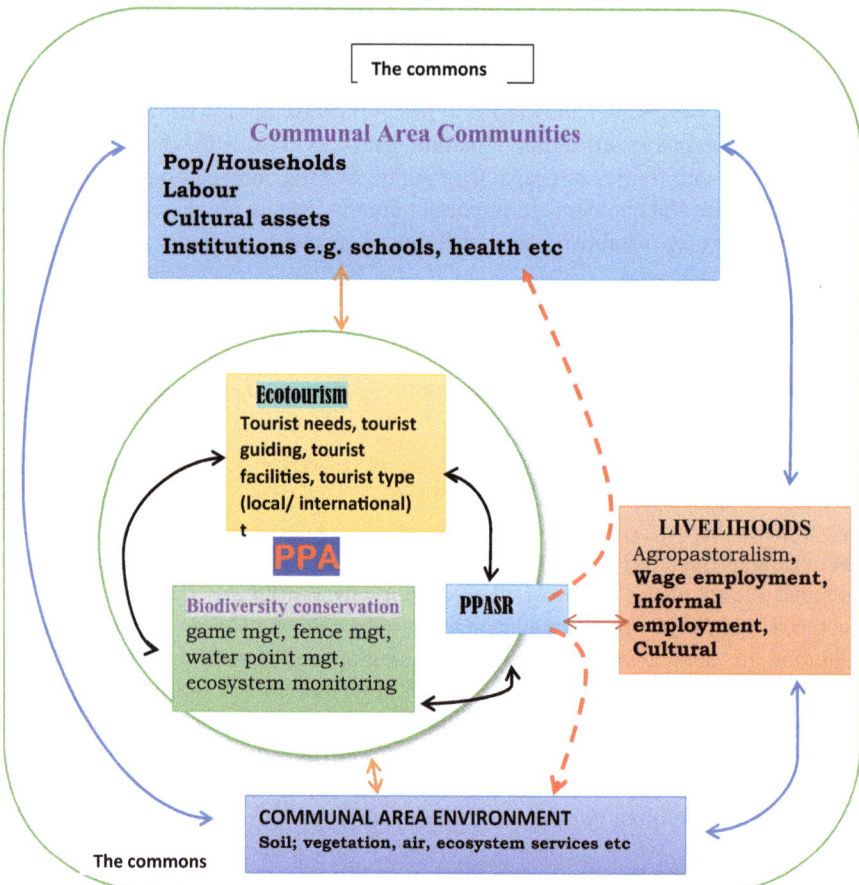

Fig. 11.1 Conceptual framework. (Source: Modified from WEFf (2012) and Bizikova et al. (2013). Key: PPASR private protected area social responsibility, PPA private protected area, and Interdependent linkage ⟷ Outward service ⟶ ⟶ ⟶)

Livelihoods. The three inner components act as cogs of the wheel interlinked to each other for the internal stability and sustainability of the PPA and its related components. PPASR reflects the kind gesture extended to the local communities. Therefore, the NT model is about promoting co-existence, tangible benefits sharing, neighbourliness and mutual relationship between the PPA and the local community to keep both the local communities and the natural ecosystem self-sustainable, efficient, self-sufficient, and curb unnecessary burden on the Private Protected Area resources.

11.3   The Study Area and Methodology

11.3.1   Mokolodi Nature Reserve

Mokolodi Nature Reserve (MNR) is a privately owned entity established in 1994 by Mokolodi Wildlife Foundation (MWF) to promote environmental education and conservation (MWF, 2010). MNR is a charitable trust registered under the Botswana Societies Act of 1972 (MWF Booklet, 2016–17). The reserve covers approximately 4500 hectares of acacia bushveld valley (MWF, 2010). Amongst all the neighbouring communities, Mokolodi community appears to command direct and occasional interactions with MNR, and this is so because a significant number of employees at MNR come from the community (MWF, 2016–17). Hence the choice of the study area (Fig. 11.2). Although Mokolodi community was formally recognised in 2006, it existed long before the establishment of MNR. Available records indicate that as early as 1933, few people were already inhabitants in the area (MWF, 2010). Today, some of the residents of this community are the second and third generation of original inhabitants. The community has a population of about 652 (CSO, 2011). The community has an assembling place known as Kgotla, administered by a headman (Kgosi) Boitshoko Rasethogwane and Village Development Committee (VDC).

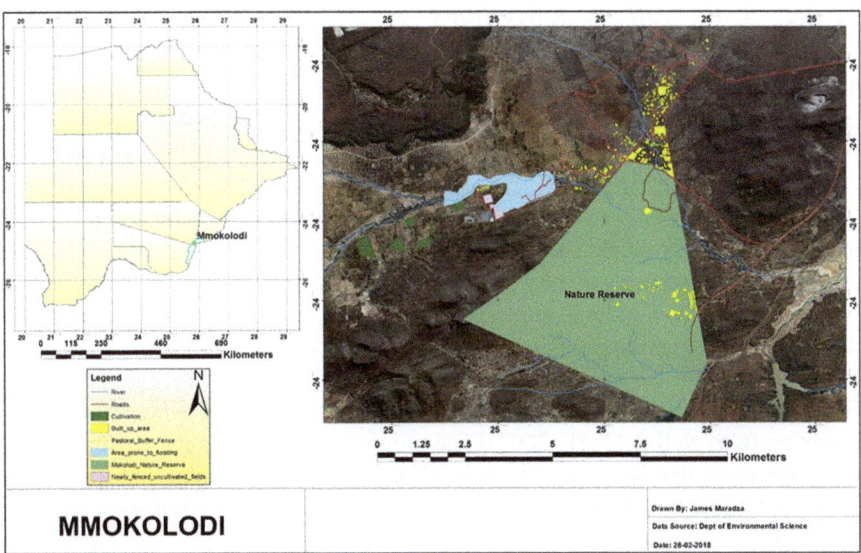

Fig. 11.2 Study area of Mokolodi Nature Reserve. (Source: Authors)

11.4 Methodology

11.4.1 Target Population and Research Design

MNR has a diversity of stakeholders (cf. Table 11.1) who are keen not only to see the MNR succeeding but also thriving in a mutually beneficial relationship between 'the Reserve' and the adjacent local community. The study gauged the views, ideas, and perceptions of stakeholders on the opportunities and challenges of adopting the NT framework. An in-depth face-to-face interview was carried out with identified informants from different stakeholder groups, whose positions, personal skills, and knowledge put them in good stead to have sufficient insights on the information sought.

The mixed-methods approach was adopted as a research framework for generating and analysing data. The mixed-methods framework consisted of two sources of field data: key informant interview reports and secondary data. The mixed-methods approach refers to the combination of at least two methods in the study of the same phenomenon (Creswell, 2003) so that the final data draws inferences from both qualitative and quantitative information. The approach helped to gain insights into the NT approach, and the sought stakeholder perceptions addressed the issue of benefits sharing and challenges of applying the NT.

11.4.2 Data Collection Methods

The study was premised on the qualitative approach to establish the perception of stakeholders on NT model and was conducted between the month of May to October in 2018. Purposive sampling and snowball techniques were used for selecting key informants (Kumar et al., 1993). The sampling procedure of key informants assumed that they are knowledgeable and can provide insight to the issues being researched and are willing to provide such information (Kumar et al., 1993) by virtue of their knowledge or experience in their field of study, for the convenience of the researcher who applies his own judgment to select the informants (Moswete & Thapa, 2018). The interviews were targeting people holding leadership positions in Government departments, Non-Governmental Organisations (NGOs) and the local community. Representatives were purposively selected from stakeholders using convenience sampling (Creswell & Clark, 2007). The rationale for targeting these officials was the assumption that because of their level of education, positions, and skills, they have been directly or indirectly involved with the operations and management initiatives of tourism, conservation, and livelihoods at MNR and the local community. In addition, leaders, by nature of their duties in respect of this area of study, are expected to be involved in matters dealing with decision making, governance and resource utilisation and management. The sample size consisted of 10 (n = −10) informants who included men and women above the age of 18. However, more than three-quarters of the total sample (n = 8) informants responded to the interview request, and their perceptions in regard to the applicability of NT model on private protected areas was solicited.

The key informant guide, which served as a checklist to guide the interview, was prepared by the interviewer and was used to solicit qualitative data as interview information. A total of eight out of ten usable key informant responses were gathered from an in-depth face-to-face interview. A probing technique was employed to elicit more information and to seek clarity from informants who freely expressed their opinions, viewpoints, conclusions and recommendations throughout the interview sessions.

11.5 Results

11.5.1 Key Informant Stakeholders

The study findings show that most of the stakeholders interviewed were mature, educated, and expertise in their field of work. This revealed that the information gathered was sought from an enlightened set of key stakeholders with arguably relevant and reliable information for the study because of their relevant education and/or experience. Although the chief and the village development committee (VDC) officer possessed minimal formal education, they, however, jointly presented reliable and relevant information about conservation, ecotourism and even the genealogy of their people and traditional lineage. The findings from the study noted a gender bias or disparity in managerial positions both in private and government institutions.

A total of eight of ten key informants were identified and interviewed (Table 11.1). The other two out of ten identified informants were not at liberty to share out their perceptions. Of the eight informants interviewed, only one was female. This gender discrepancy can partly be attributed to the non-random sampling technique used by the researcher, which gave preference only to people willing to give information by virtue of their knowledge or experience in their field of study. At least more than half of the interviewees were above 50 years of age, and a substantial number of them had attained a postgraduate qualification at the level of first degree or master's degree in their professional field, and an even higher proportion more than two-thirds had at least 10 years of experience in their field of expertise. Generally, the key stakeholders interviewed had the knowledge and the capability of responding to interview questions asked and understood exactly what the researcher wanted to achieve in the study.

Table 11.1 Key informant stakeholders' demographic profiles

Informant institution	Professional title of informant	Age	Sex	Educational qualifications	Years of experience
1	Tribal Administration	Mokolodi Village Chief	76	M	Standard 7	27
2	Kalahari Conservation Society	Chief Executive Officer	60	M	Bachelor degree	1
3	Birds Life Botswana	Program Officer	38	F	Bachelor degree	12
4	Local Government	Community Dev Officer (SEDC/VDC)	55	M	Form 2	10
5	Local Government	Principal Planning Manager (SEDC)	37	M	Bachelor degree	13
6	Central Government	Chief Natural Resources Officer (DEA)	54	M	Master degree	10
7	Central Government	Chief Tourism Officer (DoT)	45	M	Master degree	20
8	Central Government	Chief Wildlife Officer (DWNP)	55	M	Master degree	38
9	Botswana Tourism Board	Did not participate/ Refusal				
10	The Kgalagadi Beverage Trust (KBT)	Did not participate/ Refusal				

11.5.1.1 Stakeholder Perceptions on the Opportunities and Challenges of Applying Nexus Thinking: Conservation, Ecotourism and Local Livelihoods for the MNR

Poverty challenges for SDG 1 Poverty reduction is a global concern as reflected in the United Nations' SDG 1. To gauge poverty perception among members of Mokolodi Community, stakeholders were asked if poverty is a challenge in the country. All the respondents agreed that poverty is a menace in Mokolodi and among the local people in Botswana. One of the respondents even cited *the "lack of sustainable livelihood incomes as a major factor contributing to poverty in most of these rural communities"* (Stakeholder 1). When asked how poverty can be addressed in local communities that are close to privately-owned protected areas like MNR, many of the respondents suggested small scale income-generating projects. In their suggestions, the informants alleged that: *"we need small income-generating projects such as poultry, basketry, artefacts and or beekeeping"* (Stakeholder 2).

Further comments were suggested for government to focus on formulating policies that support community involvement in Private Protected Areas (PPAs) to improve local livelihoods. This view was shared by one of the respondents who affirmed that *"there should be policies that avail the local communities' appropriate programs for long term economic benefit"* (Stakeholder 5). The research also shows that only two respondents viewed PPAs as sole providers of employment to the local community. This is probably because, whilst it is important for PPAs such as MNR to create employment, they cannot realistically absorb all the people from the local community. Therefore, entrepreneurship projects linked to PPA have been perceived as having wider socio-economic benefits and complement the limited direct job opportunities.

Stakeholder perceptions of NT as a holistic approach and opportunities associated thereof The research instrument elicited informants' perceptions as to whether the approach had a desirable impact on conservation, ecotourism and livelihoods or not. All the key informants positively concurred with the assertion. In support, the majority of the informants indicated that the approach promoted interaction, collective thinking, and empowerment for the local community. One officer from the NGO sector had this to say:

> ... the local community would feel that they are directly involved and custodians of the natural resources even though the wildlife is privately owned. The approach would also promote partnerships ... enhance cooperation and good relationship (Stakeholder 7).

Two informants indicated that the NT approach would bring value that would drive the local community to conserve nature for their livelihood benefit. One officer from the government related sector concluded that NT:

> ... for there to be ecotourism, you need conservation to thrive, and ecotourism will create employment and improvement of skills. The adjacent community will see value to conserve because when people derive any livelihood benefits, they are bound to conserve natural resources (Stakeholder 3).

To further assess the concept of NT, the informants were asked if they perceived the approach as a sustainable framework. A total of 7 informants agreed that NT is a sustainable and effective framework. Most of them indicated that it is a win-win solution to poverty and the loss of biodiversity. One of the officers based in the NGO sector affirmed that *"...NT is a sustainable framework and a win-win solution to socio-economic and ecological challenges"* (Stakeholder 7). However, one Informant in the government sector was pessimistic:

> NT is something that can be looked into, studied and analysed further to pick the actual benefits out of the system. If done sustainably adjacent to PPAs where communities attain royalties, this would be a much more beneficial concept (Stakeholder 3).

Challenges in adopting the NT The study found out that there are challenges in adopting the NT approach. The majority of the stakeholders cited a possible lack of tangible benefits to the local community. The informants were of the view that *"if benefits from the concept are not tangible, members of the community may not embrace the initiative"*. Some of the Informants were of the view that most owners of PPAs were after profit and would not be willing to engage the local community as this could divert their profit-making ideology to social responsibility, which they felt, rightly, is the sole responsibility of government.

Foreseeable risks to NT adoption The study also established stakeholder perceptions of foreseeable challenges to the framework. All the informants concurred that
there were slight foreseeable challenges. The cited challenges include: change of
administration; increase in population; natural disasters such as climate change and
droughts. These challenges usually could lead to loss of biodiversity and ultimately
affect ecotourism. An upset to one component automatically affects the others since
the components are interlinked. Mismanagement in a system whose components
depend on one another may result in a total collapse. As indicated by of one of the
informants:

> Mismanagement is the greatest enemy when it comes to a system that is interlinked and
> involving sensitive environmental components: this is to say a loss or complication in one
> component will see everything collapsing… (Stakeholder 4).

Change of ownership of the PPAs was noted as a challenge in a system that relies on
interconnectedness. One of the informants from the government sector asserts that
after the owner of a PPA acquires enough profits, *"he or she may decide to quit the
business and sell the business at any given time without consulting anybody, leaving
the whole community relying on that venture in limbo"* (Stakeholder 8). The person
who buys the venture brings in his own management style, which may have little to
no interest in the needs of the local community.

Stakeholder Recommendations for adopting NT: Finally, the informants were
asked if they would recommend the NT framework. All the informants strongly
recommended the adoption of the novel NT framework. The majority of the informants cited benefits such as empowerment, improvement of conservation, ecotourism and livelihoods of the local community. As summed up by one of the informants
from the government sector, *"Yes, I recommend the adoption of the concept because
it is a holistic approach and a win-win solution to the socio-economic and ecological challenges"* (Stakeholders 3). Another respondent from the private sector also
weighed in:

> I strongly recommend the approach because you need support from your surrounding com-
> munity if you are to be successful in conservation initiatives, and the support is only guar-
> anteed if and when the local community realises benefits (Stakeholder 1).

The informants felt that the approach has prospects of being successful and sustainable, provided possible challenges anticipated are strategically identified and
mitigated.

11.6 Discussion, Conclusion and Recommendations

Generally, the stakeholders portrayed a positive perception of the applicability of
Nexus Thinking (NT) to Private Protected Area (PPAs) and Government Protected
Areas (GPAs). This finding is logical, considering that the NT model is built upon
interaction, collaboration, entrepreneurship (see Bazilian et al., 2011; Mbaiwa et al.,
2011; WEF 2012). More so, informants affirm that the NT approach is a win-win

solution to socio-economic and ecological challenges. However, some of the stakeholders were pessimistic. They felt that NT was rather abstract and further assessment was required to refine it. Nonetheless, the majority of informants was optimistic and viewed the framework as noble, practical and interactive.

Largely, the results of this study portray NT as a model applicable to conservation, ecotourism and livelihoods in PPAs and surrounding communities. The literature review conclusively reveals that the NT approach is a sustainable framework linked by multiple interacting components (Bazilian et al., 2011; WEF 2012). The study acknowledges the strength of the NT not only in dealing with the socio-economic and ecological challenges (e.g. poverty, loss of biodiversity and environmental degradation) but also in guaranteeing socio-economic and ecological opportunities (e.g. employment, entrepreneurship, social amenities), benefiting both PPAs and proximate communities.

This framework enhances our understanding of the concept of NT model and how different components interdepend with each other. Notably, it is this integration (linkage) that keeps the components sustainable. It is with no doubt that the linkages have the potential to unlock conservation, ecotourism and livelihoods enterprises and transform Batswana's livelihoods. The study adds literature on NT and its application to PPAs and GPAs, which is currently limited. The study afforded environmental planners, government, stakeholders and entrepreneurs' insight into the potential of NT in addressing the loss of biodiversity, poverty and environmental degradation. Generally, based on the study findings, the NT model reflected an innovative way of addressing both human and environmental challenges and building sustainable interconnected human-wildlife societies and ecosystems.

The study findings are based on empirical evidence from the study, which show that NT is an interactive framework that promotes linkages of different components. Therefore, the study recommends that the Local village and MNR engage in formal interactive partnership agreements that enhance local village investment and entrepreneurship. The study findings further reveal that corporate social responsibility promotes interaction and good relationships between the adjacent local villages and the MNR. That is, MNR may have to develop a local village empowerment programme particularly targeting the less privileged – the youth and the women – and equip them with entrepreneurial skills to improve their livelihoods.

References

Adams, W. M., Aveling, R., Brockington, D., Dickson, B., Elliott, J., Hutton, J., & Wolmer, W. (2004). Biodiversity conservation and the eradication of poverty. *Science, 306*(5699), 1146–1149.

Armitage, D. (2005). Adaptive capacity and community-based natural resource management. *Environmental Management, 35*(6), 703–715.

Arntzen, J., Buzwani, B., Setlhogile, T., Kgathi, D., & Motsolapheko, M. (2007). *Community-based resource management, rural livelihoods and environmental sustainability.* Centre for Applied Research.

Bazilian, M., Rogner, H., Howells, M., Hermann, S., Arent, D., Gielen, D., & Tol, R. S. (2011). Considering the energy, water and food nexus: Towards an integrated modelling approach. *Energy Policy, 39*(12), 7896–7906.

Benson, D., Gain, A., & Rouillard, J. (2015). Water governance in a comparative perspective: From IWRM to a 'nexus' approach? *Water Alternatives, 8*(1), 753–773.

Beresford, M., & Phillips, A. (2000). Landscape stewardship: New directions in conservation of nature and culture. *The George Wright Forum, 17*(1), 15–26.

Bizikova, L., Roy, D., Swanson, D., Venema, H. D., & McCandless, M. (2013). *The water-energy-food security nexus: Towards a practical planning and decision-support framework for landscape investment and risk management*. International Institute for Sustainable Development.

Blaikie, P. (2006). Is small really beautiful? Community-based natural resource management in Malawi and Botswana. *World Development, 34*(11), 1942–1957.

CSO. (2011). *2011 Botswana population and housing census*. Statistics Botswana.

Central Statistics Office (CSO). (2018). Population and housing census 2011: Analytical report. www.cso.gov.bw/cso

Christian, M., Fernandez-Stark, K., Ahmed, G., & Gereffi, G. (2011). The tourism global value chain: Economic upgrading and workforce development. *Skills for Upgrading*, 276–280.

Creswell, J. W. (2003). *Research design: Qualitative, quantitative, and mixed-method approaches* (2nd ed.). Sage Publications.

Creswell, J. W., & Clark, V. P. (2007). *Designing and conducting mixed methods research*. Sage Publishers.

Crotti, R., & Misrahi, T. (2017). *The travel and tourism competitiveness report 2017: Paving the way for a more sustainable and inclusive future*. World Economic Forum.

Culpeper, R. (2005). *Approaches to globalisation and inequality within the international system*. United Nations Research Institute for Social Development.

DeGeorges, P. A., & Reilly, B. K. (2009). The realities of community-based natural resource management and biodiversity conservation in Sub-Saharan Africa. *Sustainability, 1*, 734–788.

Government of Botswana (GoB). (2001). *Botswana national atlas*. Government Press.

Government of Botswana (GoB). (2007). *Community based natural resources management policy. Government paper N0. 2 of 2007*. Government Press.

Groenfeldt, D. (2010). The next nexus: Environmental ethics, water management and climate change. *Water Alternatives, 3*(3), 575.

Fabricius, C., & Collins, S. (2007). Community based natural resources management governing the commons. *Water Policy, 9*, 83–97.

Hoole, A. F. (2008). *Community-based conservation and protected areas in Namibia: Social-ecological linkages for biodiversity*. Thesis, University of Manitoba, USA.

Hussey, K., & Pittock, J. (2012). The energy-water nexus: Managing the links between energy and water for a sustainable future. *Ecology and Society, 17*(1), 31. http://www.ecologyandsociety.org/vol17/iss1/art31

Kumar, N., Stern, L. W., & Anderson, J. (1993). Conducting interorganisational research using key informants. *Academy of Management Journal, 36*(6), 1633–1651.

Laderchi, C. R., Saith, R., & Stewart, F. (2003). Does it matter that we do not agree on the definition of poverty? A comparison of four approaches. *Oxford Development Studies, 31*(3), 243–274.

Leck, H., Conway, D., Bradshaw, M., & Rees, J. (2015). Tracing the water–energy–food nexus: Description, theory and practice. *Geography Compass, 9*(8), 445–460.

Matseketsa, G., Chibememe, G., Muboko, N., Gandiwa, E., & Takarinda, K. (2018). Towards an understanding of conservation-based costs, benefits, and attitudes of local people living adjacent to Save Valley Conservancy, Zimbabwe. *Scientifica*. https://doi.org/10.1155/2018/6741439

Mbaiwa, J. E. (2011). The effects of tourism development on the sustainable utilisation of natural resources in the Okavango Delta, Botswana. *Current Issues in Tourism, 14*(3), 251–273.

Mbaiwa, J. E., Stronza, A., & Kreuter, U. (2011). From collaboration to conservation: Insights from the Okavango Delta, Botswana. *Society and Natural Resources, 24*, 400–411.

Mbaiwa, J. E., Mbaiwa, T., & Siphambe, G. (2019). The community-based natural resource management programme in southern Africa–promise or peril? The case of Botswana. In M. Mkono (Ed.), *Positive tourism in Africa* (pp. 11–22). Routledge.

Mokolodi Wildlife Foundation. (2010). *Mokolodi: Wild dogs visits Mokolodi, local fires test Mokolodi differences*. Mokolodi Nature Reserve.

Mokolodi Wildlife Foundation. (2016). *Mokolodi 2016–17 annual review: Adapting in maturity*. Mokolodi Nature Reserve.

Mopelwa, G., & Blignaut, J. (2014). The Okavango Delta: The value of tourism. *South African Journal of Economic and Management Sciences, 9*(1), 113–127.

Moswete, N., & Thapa, B. (2018). Local communities, CBOs/trusts, and People–Park relationships: A case study of the Kgalagadi Transfrontier Park, Botswana. *The George Wright Forum, 35*(1), 96–108.

Murphree, M. W. (1998). *Congruent objectives, competing interests and strategic compromise: Concepts and process in the evolution of Zimbabwe's CAMPFIRE Programme*: Institute for Development Policy and Management, University of Manchester, Helen, Georgia, USA.

Mutandwa, E., & Gadzirayi, C. T. (2007). Impact of community-based approaches to wildlife management: Case study of the CAMPFIRE programme in Zimbabwe. *The International Journal of Sustainable Development & World Ecology, 14*(4), 336–344.

Nyaupane, G. P., & Poudel, S. (2011). Linkages among biodiversity, livelihood, and tourism. *Annals of Tourism Research, 38*(4), 1344–1366.

Phuthego, T., & Chanda, R. (2004). Traditional ecological knowledge and community-based natural resource management: Lessons from a Botswana wildlife management area. *Applied Geography, 24*(1), 57–76.

Pimm, S. L., Jenkins, C. N., Abell, R., Brooks, T. M., Gittleman, J. L., Joppa, L. N., & Sexton, J. O. (2014). The biodiversity of species and their rates of extinction, distribution, and protection. *Science, 344*(6187), 124–143.

Rabaloi, B. (2006). *Potential of tourism for economic development in Bostwana: And application for SAM multiplier analysis*. Unpublished, MA Thesis, Department of Economics, University of Botswana.

Robertson, I. (1989). *Society: A brief introduction*. Worth-Publishers.

Schlossberg, S., Chase, M. J., & Sutcliffe, R. (2019). Evidence of a growing elephant poaching problem in Botswana. *Current Biology Report, 29*, 2222–2228.

Scoons, I. (2009). Livelihoods perspectives and rural development. *Journal of Peasant Studies, 36*, 171–296.

Sebele, L. S. (2010). Community-based tourism ventures, benefits and challenges: Khama rhino sanctuary trust, central district, Botswana. *Tourism Management, 31*(1), 136–146.

Snyman, S. (2013). Household spending patterns and flow ecotourism income into communities around Liwonde National Park, Malawi. *Development Southern Africa, 30*(4–5), 640–658.

Steiner, A., & Rihoy, E. (1995). The commons without a tragedy? Strategies for community based natural resource management in Southern Africa. In *A review of lessons and experiences from natural resources management programme in Botswana, Namibia, Zambia and Zimbabwe*. Lusaka.

Stone, M. T. (2013). *Protected areas, tourism and rural community livelihoods in Botswana*. Arizona State University.

Suich, H., Howe, C., & Mace, G. (2015). Ecosystem services and poverty alleviation: A review of the empirical links. *Ecosystem Services, 12*, 137–147.

Swatuk, L. A. (2005). From "project" to "context": Community-based naural resource management in Botswana. *Global Environmental polotics, 5*(3), 95–119.

UNWTO (United Nations World Tourism Organization). (2018). Tourism highlights. e-unwto.org. Accessed Jan 2022.

Water-Energy-Food. (2012). *Water security: The water-food-energy-climate nexus*. Island Press.

WTTC (World Travel & Tourism Council). (2009). *Travel & tourism economic impact: Executive, summary, 2009*. London, UK. Retrieved from: http://www.wttc.org/bin/pdf/original_pdf_file/exec_summary_2009.pdf

World Bank. (2020). *The World Bank in Botswana*. The World Bank partners with the government to promote private sector-led, jobs-intensive growth, strengthen human and physical assets, and support effective resource management. Retrieved from: https://www.worldbank.org/en/country/botswana/overview

James Maradza has a Master of Science, Environmental Science from the University of Botswana. His research interests are on the challenges of environmental degradation, biodiversity loss and climate change: exploring opportunities collaborating on environmental issues in ways that interlink people with nature for sustainable development in southern Africa.

Raban Chanda is a Human Geographer with long teaching and research experience in Botswana and Zambia. His area of expertise includes environmental assessments, sustainable livelihoods, socio-economic dimensions of environmental change and natural resource management, and climate change adaptation. He is a Full Professor in Environmental Science at the University of Botswana.

Naomi N. Moswete is a Senior Lecturer in the Department of Environmental Science, University of Botswana. Her research interests include Human geography, tourism as a strategy for rural development, community-based tourism; Transboundary conservation areas –ecotourism nexus, parks–people relationships, heritage management & cultural tourism.

Chapter 12
Environmental Change, Wildlife-Based Tourism and Sustainability in Chobe National Park, Botswana

Maduo O. Mpolokang, Jeremy S. Perkins, Jarkko Saarinen, and Naomi N. Moswete

12.1 Introduction

Wildlife-based tourism remains a dominant activity and an important part of the tourism industry in southern Africa. Wildlands and wilderness environments have become a drawcard to millions of visitors, especially to the region's conservation areas and game reserves (Moswete et al., 2017; Shoo & Sorongwa, 2013; Spenceley, 2008). This makes wildlife-based tourism big business, generating large revenues for governments, foreign exchange earnings and jobs, especially in rural areas suffering from high unemployment rates (see Dobson, 2006; Earnshaw & Emerton, 2000; Mbaiwa, 2018; Santarém et al., 2018). In Botswana, tourism contributes over 10% of the Gross Domestic Product (GDP) (WTTC, 2015, 2018), and the industry relies on wildlife and nature-based tourism, in general, which accounts for almost 90% of the tourist revenue (DWNP, 2008).

Botswana is recognised globally as a premier destination for viewing wildlife and wilderness environments (Mbaiwa, 2017; Mogende & Moswete, 2018), and the promotion of tourism has been increasingly used for economic diversification (Saarinen et al., 2012; UNWTO, 2008). As a result, wildlife-based tourism has grown, and an increasing number of communities have been integrated into emerging economic opportunities based on tourist consumption (Moswete & Thapa, 2015,

M. O. Mpolokang · J. S. Perkins · N. N. Moswete (✉)
Department of Environmental Science, University of Botswana, Gaborone, Botswana
e-mail: perkinsjs@ub.ac.bw; moatshen@ub.ac.bw

J. Saarinen
Geography Research Unit, University of Oulu, Oulu, Finland

School of Tourism and Hospitality, University of Johannesburg, Johannesburg, South Africa
e-mail: jarkko.saarinen@oulu.fi

© The Author(s), under exclusive license to Springer Nature   169
Switzerland AG 2022
J. Saarinen et al. (eds.), *Southern African Perspectives on Sustainable Tourism Management*, Geographies of Tourism and Global Change,
https://doi.org/10.1007/978-3-030-99435-8_12

2018). Currently, the country's tourism is largely based on the conservation areas located in the northern parts of Botswana (GoB, 2001). The Okavango Delta, Chobe National Park and Moremi Game Reserve are the main tourist destinations, accounting for over 90% of the tourist visitations in the northern protected areas (DWNP, 2008; Mmopelwa & Mackenzie, 2020). Mobile safari operators with licences to operate in these areas have increased drastically in recent years (Mogende & Moswete, 2018; Perkins, personal observation), spreading the economic benefits of tourism more widely but also placing unprecedented pressure on the environment. These areas are also vulnerable to the impacts of global climate change (Hambira et al., 2013, 2020; Saarinen et al., 2022), which causes a myriad of challenges to wildlife-based tourism relying on the conditions of natural environment and ecosystem processes (Lepetu & Garekae, 2020; Mogende & Moswete, 2018; Moswete et al., 2017).

Threats to wildlife-based tourism are primarily driven by both natural and human factors such as habitat loss or fragmentation, unprecedented climate change, veld fires and the burgeoning elephant population (Kilungu et al., 2019; Santarém et al., 2018). Environmental change, together with some of its drivers, has considerable potential to negatively affect the country's sustainability with regard to wildlife-based tourism (Mkiramweni, 2014; Mogende & Moswete, 2018; Nyaupane & Chhetri, 2009). Despite clear implications from environmental change connected with the global warming crisis (Engelbrecht et al., 2015), fires (Pricope & Binford, 2012; Fox et al., 2017), ecosystem fragmentation (Naidoo et al., 2018), wildlife-based tourism in northern Botswana has received limited research attention especially in the context of sustainability management under environmental change.

Conceptually, wildlife safaris or wildlife-based tourism includes non-consumptive and consumptive forms (Fennell, 1999; Mabunda & Wilson, 2009). In general, the former refers to interactions with wildlife that are based on observing and photographing animals in their natural habitats. The latter, i.e. consumptive wildlife tourism, includes activities such as fishing and sport or trophy hunting in which wildlife is merely seen as a resource for human consumption (see Campbell, 2008; Lovelock, 2008; Novelli et al., 2006; Santarém et al., 2019). According to Spenceley (2008), wildlife tourism becomes consumptive when wildlife is killed. However, wildlife viewing in captive or semi-captive situations can also be included as wildlife tourism (Newsome et al., 2005, p. ix), and it is debatable whether such forms of wildlife tourism are truly non-consumptive. Rather, the relationship between non-consumptive and consumptive wildlife tourism is not clear-cut but a continuum and potentially complex.

In this chapter, we focus on the non-consumptive wildlife-based tourism that takes place in nature conservation areas, particularly in the Chobe National Park (CNP), Botswana. According to Barnes (2001), a non-consumptive wildlife-based tourism should be prioritised in the southern African context. This should be planned and developed in a sustainable way (Saarinen et al., 2020; Snyman & Spenceley, 2012) that benefits local communities that often bear a burden of living and practising their livelihoods with wildlife populations (Child & Barnes, 2010; Chiutsi & Saarinen, 2017). Therefore, social, economic and environmental justice are integrally linked to sustainability thinking in wildlife-based tourism (Nyirenda et al., 2020; Saarinen, 2014). Sustainable tourism, in general, refers to *tourism that meets the needs of present and host regions while protecting and enhancing opportunities for the future* (see Liu, 2003, p. 460; Reddy & Wilkes, 2013, p. 3). It is envisaged as leading to the management of all resources in such a way that economic, social and aesthetic needs can be fulfilled while maintaining cultural integrity, essential ecological processes, biological diversity and life support systems. The theory is informed by economic, social, essential ecological processes, and biological diversity and life support systems (Liu, 2003). Therefore, sustainable tourism development is appealing as a framework to assess wildlife based-tourism, as an environmental change in vegetation cover and wild animals, drought, floods and erratic rainfall are envisaged to affect tourism sustainability (see Chilembwe, 2020; Mogende & Moswete, 2018; Mosugelo et al., 2002 Perkins, 2019).

This paper discusses the potential threats posed by environmental change on wildlife-based tourism towards sustainability in Chobe National Park (CNP), Botswana. The study focuses on the perceived and estimated impacts of environmental change to wildlife tourism in CNP. The study uses a mixed-method approach based on a survey for safari tour guides and interviews on key stakeholders from different governmental offices and representatives of non-governmental organisations (NGOs) working with wildlife-based tourism policies and development in Botswana.

12.2 Methodology

12.2.1 The Case Study Area

The study was conducted in the northern part of Botswana in the Chobe National Park in Chobe District (Fig. 12.1). The Park was established in 1961 and is the second-largest in the country. It covers approximately 10,590 km^2 (DWNP, 2008; GoB, 2001), which consists of floodplains, swamps and woodland (BTO, 2016). The Park supports a unique diversity and concentration of wildlife, has one of the largest concentrations of fauna and flora in Africa and is one of the largest parks in the country (BTO, 2016). Some of the main economic activities in the area include crop production, livestock production and wage employment. These economic activities are complemented by small-scale businesses such as selling baskets, game meat and thatching grass (Jones, 2002).

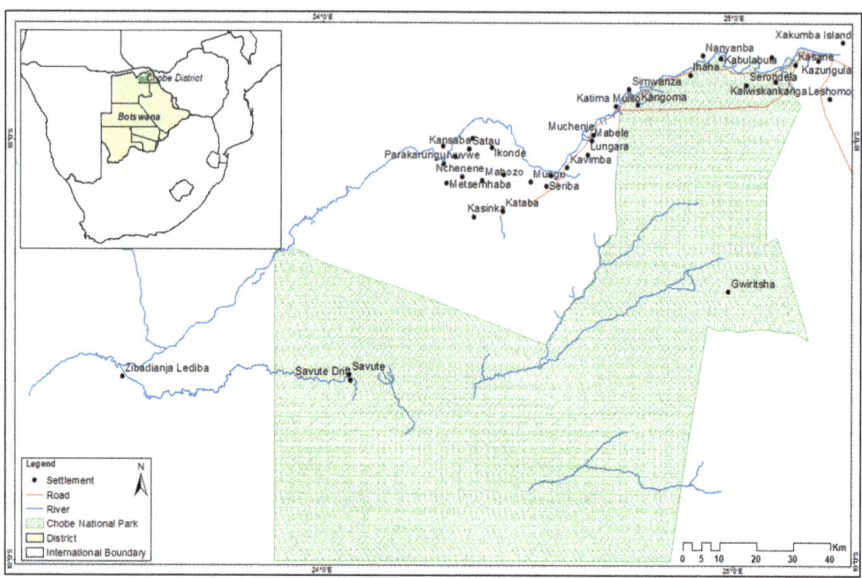

Fig. 12.1 Map showing the Chobe National Park in northern Botswana

12.2.2   *Research Materials and Analysis*

The paper employed a mixed-method approach (Creswell, 2013), which was chosen for reasons that it has a capacity to improve the usability of the collected data by providing combined information from complementary kinds of data and having an ability to integrate divergent views and make stronger inferences. In respect to research materials, the study draws largely from fixed and mobile safari tour guides, government officials and (NGOs), especially those who work on environmental change matters, in and around CNP, as the target population. The study surveyed 63 mobile and fixed tour guides who were purposively sampled from the provided list of safari lodge operators. The license for wildlife tour guides allows for both mobile and fixed guiding. Therefore, it was not possible to separate mobile tour guides from the fixed tour guides, hence they were treated the same during data collection. In addition, nine key informants were also purposively sampled. In the context of this study, government officials, Non-Governmental Organisations (NGOs), conservation-based organisations, and wildlife tour guides were considered carefully based on the author's prior knowledge of their responsibilities, knowledge and participation in conservation programmes. Key informants included representatives from; Department of Wildlife and National Parks (DWNP), Elephants Without Borders (EWB), Botswana Tourism Organisation (BTO), Centre for Conservation of African Resources, Animals, Communities and Land use (CARACAL), Kavango-Zambezi Transfrontier Conservation Area (KAZA), Department of Forestry and Range Resources (DFRR), Land board.

Self-administered semi-structured questionnaires were used to collect data from wildlife tour guides representatives. A semi-structured interview through a semi-structured questionnaire was used to gather focused qualitative and quantitative data, experiences and perceptions of the mobile and fixed wildlife tour guides. The questionnaires were administered to the respondents through face-to-face interviews, conducted by the researcher at the interviewees' place of work and at their convenient times. This was to allow the researcher to explain and clarify the questions where necessary. The questionnaire gathered demographic and socio-economic information of the respondents; nature and occurrence of environmental events; potential impacts of environmental change on wildlife-based tourism. An interview guide was used to solicit information from the key informants to provide wildlife resources information, fire history and vegetation cover information and the tourist numbers in CNP and technical knowledge on tourism issues. Secondary data in the form of academic publications, unpublished documents, consultant reports and management plans were used. All ethical procedures were followed. The University of Botswana Policy on Ethics and Ethical Conduct in Research, policy reference number RD 04/05H was taken into consideration. The completed application form from the Office of Research and Development (ORD) was submitted to the Ministry of Environment, Natural Resources Conservation and Tourism for approval and to acquire the permit. Informed consent was respected; the researcher ensured that each participant's information was kept anonymous and confidential from any other person. No participant was forced to participate in the study, hence it was completely voluntary.

Data was compiled and managed using Statistical Package for Social Sciences (SPSS) version 25. Descriptive statistics such as frequencies and means were used to analyse quantitative data, while content analysis was used for cleaning and organising qualitative data. In this study, data was organised based on the recurring themes after grouping.

12.3 Results and Discussion

12.3.1 Demographic Profile

A total of 63 safari tour guides participated in the study. About 76.2% (n = 48) reside in Kasane while 23.8% (n = 15) in Kazungula. The majority of the tour guides were males (98.4%, n = 62), but one was female. This gender difference could be attributed to cultural notions since a safari tour guide job is perceived to be a male job, hence their dominance. Literacy levels varied amongst the safari tour guides, and the levels had vast differences across the study population. The predominant education level among the tour guides was at the secondary level (47.6%, n = 30) while university degree constituted the least (7.9%, n = 5), 27.0% had attained technical/vocational certificates. Only 17.5% of the respondents noted others such as Dreams Safaris and Botswana Wildlife Training Institution (BWTI), which offer tour guiding certificates. The mean length of years worked as a tour guide was 7.94. This indicates that most of the tour guides worked for a few years for less than 5 years (52%) as tour guides.

Key informant representatives included expertise of diverse backgrounds from non-governmental organisations and government authorities such as; a Tourism Development Manager (BTO), Research Officer (DWNP), Problem Animal Control (DWNP), Park Manager (DWNP), Programme Officer (EWB), Chief Executive Officer (CARACAL), Land Board (Department of Town and Country Planning) Board (Senior Lands Officer), Liaison Officer (KAZA) and research officer (DFRR). These representatives were asked to give information relating to environmental change and knowledge, tourism and conservation issues, stakeholder involvement and regulation of local products and/or resources. Attempts were made to contact potential respondents in advance through scheduling appointments with potential respondents. The interviews took approximately between 30 and 45 minutes and were conducted in the respondent's respective administration offices.

12.3.2 Drivers of Environmental Change

The results of this study highlight the complex drivers of environmental change and how they affect the environment. Generally, results show that the study area as a whole experiences an assortment of drivers, especially those that are due to natural factors. The perceived natural factors driving environmental change in the study area include drought, fires, increased temperatures, and changes in surface water availability – flooding and cessation of flows in river and lack of surface water in surface depressions or pans in the wet season due to droughts, diseases and burgeoning elephant population. All of these factors increase environmental change costs to the wildlife and tourism systems. To a large degree, they apply to CNP and explain why environmental issues cause concern. However, for this study, according to some key informants, there are some anthropogenic activities that threaten the environment. The anthropogenic factors influencing environmental change include deforestation and congestion of park users in the Park, which causes soil erosion. Key informants indicated that it deprives the Park of many things. For instance, it affects some species such as the sable antelope (*Hippotragus niger*) and scares away shy species such as the Chobe bush-buck (*Tragelaphus scriptus*), and similarly affects and influence vegetation change, especially along the Chobe River Front (CRF). The eastern Chobe Riverfront, especially the easternmost 16kms, has for several decades suffered from tourism congestion due to the tendency for boats and game drive vehicles to crowd into this area and neglect other areas of the Park. Rules regarding the direction of travel, registering the numbers of boats and tourists entering the Park, as well as limits on vehicle numbers during peak times, have all been implemented with some success. However, the problem is set to escalate when the new road/rail bridge that links Kazungula to its neighbours opens in 2021 and delivers large numbers of tourists to the CNP entry points.

The results of this study confirm observations made by previous studies across the world, which note that climate-related factors usually act as drivers of environmental change (Cumming, 2008; Dillimono & Dickinson, 2015; Engelbrecht et al., 2015). Southern Africa will become drier and hotter due to climate change (Engelbrecht et al., 2015; Harper et al., 2018; van Wilgen et al., 2016), which will in turn impact on the multiple factors driving environmental change. The Bio-Chobe (2016) report emphasise that many natural factors such as drought and fire are set to become much more severe under anthropogenic climate change (global warming). Mosugelo et al. (2002) and Wolf (2009) have also observed that the CRF has been heavily impacted, especially by large mammals such as elephants, and there is a decline in grass cover and woody species in the CNP. Along with the congestion caused by game viewing boats and vehicles, these factors are likely to create dissatisfaction amongst tourists concerning the quality of the game viewing experience on offer as well as concern over the environmental impacts of the tourism industry – for example, degradation of the roads, river pollution, presence of invasive and alien species, as well as increasing interference/disruption to the behaviour of the animals themselves. A Code of Conduct for boats on the Chobe River was recently developed (DWNP, 2008; Mogende & Moswete, 2018), although as demand for the exceptional game viewing experience on offer continues to increase and new operators enter the market, the challenges will clearly only increase.

12.3.3   Nature of Environmental Change in CNP

Tour guides were asked about the events attributed to environmental change in the Chobe National Park (CNP). Nearly all tour guides (82.5%) indicated that there are events attributed to environmental change in the study area, while (12.7%) expressed that there have been no events experienced linked to environmental change. Also, approximately 4.8% indicated that they did not know if there has been events linked to environmental change. Some of the commonest events observed were congestion and/or crowding of tourist vehicles (55.6%), increased elephant population (36.5%) and increased temperatures (36.5%) as the most experienced events within the CNP, which contribute highly to environmental change.

Over three quarters (82.5%) of the tour guides demonstrated a high level of familiarity with the nature and extent of events associated with environmental change. Tour guides indicated events associated with environmental change such as; fires, increase in elephant population, flooding, diseases, increased temperatures. A relationship between event experienced and the occurrence of event was established through cross-tabulation. On this subject, each of the respondents indicated the occurrence of the provided linked to environmental events. The results show a variance of the occurrence of events (see Table 12.1). This is surprising and interesting as the majority of respondents indicated congestion/crowding of tour operators and

Table 12.1 Cross-tabulation of events experienced and occurrence of events

	Occurrence of event		
Event experienced	Past N (%)	Recurrent N (%)	New N (%)
Increase in elephant population	–	10 (15.9)	11 (17.5)
Congestion/crowding	9 (14.3)	12 (19.0)	14 (22.2)
Increased temperature	2 (3.2)	9 (14.3)	13 (20.6)
Diseases	1 (1.6)	7 (11.1)	–
Fire	2 (3.2)	7 (11.1)	2 (3.2)
Flooding	–	7 (11.1)	–

self-drive tourists by the CNP riverfront as a past (14.3%) recurrent (19%) and new (22.2%) event. The results also revealed that 11.1% perceive flooding to be a recurring event. While some of the key informants backed this up by stating that heavy flooding in Chobe is experienced at least every 5 years. Corresponding with the foregoing, the predominant occurrence of events experienced in the study area are displayed in Table 12.1. Results show that the events are mainly new and recurring.

Essentially, these findings concur with other studies that have also shown that there are noticed events attributable to environmental change, particularly in CNP (Bio-Chobe, 2016; DWNP, 2008; Wolf, 2009). For instance, Moswete et al. (2017) observed that tour guides from fixed and mobile lodges were aware of some factors leading to environmental change in CNP such as recurring droughts. Given the predictions of IPCC models, southern Africa in general, and Botswana in particular, will be severely affected by climate change. It will become drier and hotter, with 'megadroughts and a dramatic increase in the number of heatwaves experienced (Engelbrecht et al., 2015). In addition, Cumming (2008) stresses an increasing elephant population which will increase Human-Elephant Conflict (HEC) in northern Botswana. Indeed, the Bio-Chobe (2016) report emphasised that CNP was about to experience the 'perfect storm' caused by the convergence of climate change, increased fires, its burgeoning elephant population and the ever-growing tourism demand, accentuated further by the opening of the new bridge at Kazungula. Therefore, the findings of this study express concern that the CNP is about to face unprecedented threats to the sustainability of its tourism that will severely challenge the current management of the Protected Area.

12.3.4 Perceptions of Environmental Change on Wildlife-Based Tourism

The findings that emerged from both wildlife tour guides and key informants provided notable insights on perceptions towards environmental change. Tour guides harboured negative perceptions towards environmental change. Despite that, CNP tourists are interested in seeing free-roaming animals in the park and not necessarily the state of the environment, at least at the moment. This is not surprising because tourist numbers are showing signs of increase, although this could be in part due to increased marketing through social media and word of mouth. Although tourists are interested in encountering the number and variety of wild animals, change of seasons is also affected by environmental change which might also affect their satisfaction. Subsequently, visitors on a short stay may not realise how the climate is changing. Changes to rainfall, river flow regimes and the increased severity of fires may well change wildlife movements in the area. Fire-ravaged savannah and drought-related or disease-related die-offs of wild ungulates are understandably not popular with the majority of visitors. Similarly, the negative perceptions towards environmental change included its reduced attractiveness to tourists, death and outmigration of wildlife so, compromising the wildlife-based tourism system. Such perceptions were mainly attributed to the perceived environmental degradation in CNP. For Chobe National Park, environmental change reduces the aesthetic value of the area especially in some prime areas such as the CRF and the Sedudu Island. Furthermore, it leads to change in the species composition, such as an increase in elephants which heavily impact upon riverine vegetation and affects species such as the Red Lechwe (*Kobus leche*), the Puku (*Kobus vardonii*) and the Chobe Bush Buck (*Tragelaphus scriptus*).

However, although they hold a negative view about environmental change on wildlife-based tourism, at least some tour guides believed that environmental change would bring positive effects, especially on wildlife-based tourism growth and competitiveness in CNP. For instance, they believe that fires contribute to balancing the ecosystem, therefore tour guides expect an increase in tourist numbers given the known fact that fires are good for palatable grass suitable for wildlife, and also an increase in elephant population. Generally, the extracts from the key informants imply that environmental change will negatively influence the future tourist flows. For instance, a Tourism Development Manager who represented BTO noted:

> There are certain species of wildlife which we used to see in large numbers. For example, the Chobe waterbuck and the common duiker. We might lose some species due to veld fires, and if you look at the entire district (Chobe) from Pandamatenga you would see a change in vegetation. There is that stunted growth of vegetation due to the frequent fires, and the damage is visible. Therefore, all these lead to less attraction of tourists hence affecting wildlife-based tourism.

In addition, a DWNP officer indicated the negative influence of environmental change on tourist flows. He said:

> The amount of revenue generation will be very low, therefore reduction in the GDP because it contributes about 5% of the GDP from the collected park fees. And the other thing will be the change in the species composition like we have a lot of elephants here in the Chobe National Park, which to some extent destroy the vegetation, affecting small mammals like the Red Lechwe, the Puku and Chobe bushbuck, so there will be sort of like a barrier or some degradation.

Therefore, if not managed properly, it is likely to adversely affect the CNP by reducing the number of tourists, especially during the dry season, which has always been characterised by large numbers of wildlife. Consequent to that, it should be noted that the results imply that environmental change in CNP is likely to affect the economic sustainability of the Park. For example, the BTO Tourism Development Manager expressed that: *For sustainability, there needs to be a balance of the social, economic and environment by involving the local community.* Furthermore, EWB Programme Manager explained that: *Environmental change will affect the economic sustainability of Chobe National Park, especially when there is environmental degradation, the tourism sector will go down. Therefore, there is a need to diversify the tourism in the area by utilising the Chobe Forest Reserves, including horse-riding safaris, cultural stuff in the enclave and open the Nogatshaa area.*

In CNP, environmental change affects the consumptive value, which in turn affects tourists' satisfaction, that is, the variety and number of animals that tourists want to see is directly affected by environmental change (Mogende & Moswete, 2018). These findings are consistent with studies from elsewhere (Saarinen et al., 2012; Scott et al., 2012). Contrary to that, Preston-Whyte and Watson (2005) signalled the positive impacts of environmental change in that it can improve game viewing and attract more species of elephants (*Loxodonta africana*), bushbuck (*Tragelaphus sylvaticus*), and buffaloes (*Syncerus caffer*) which is likely to attract a lot of tourists. On the same note, Desanker and Magadza (2001) reveal that at least environmental change poses a more favourable environment for animals such as the eland (*Taurotragus oryx*), kudu (*Tragelaphus strepsiceros*), giraffe (*Genus Giraffa*), which might influence tourists to still travel especially to drier areas. Overall, perceptions of wildlife tour guides are similar to those observed from key informants' interviews. For example, the respondents hold negative perceptions towards environmental change, particularly in the study area.

12.4 Sustainability Management Issues in Wildlife-Based Tourism

The Government of Botswana recognised environmental change and introduced strategies to manage wildlife-based tourism activities through the HCLV Policy position and the decongestion strategy. These strategies emerged from the reality that there is a need for sustainable use of the natural resources for future use, especially for visitors seeking wildlife experience and lifestyles in the CNP (DWNP, 2008). Other strategies recognised to intensify environmental consciousness and responsiveness largely associated with benefitting the CNP as noted from the study results included Artificial Water Points (AWPs), provision of wildlife corridors and control of fires via buffer zones as the vital adaptation measures in place in trying to reduce environmental change in CNP.

The provision of AWPs in protected areas (Owen-Smith, 1996), once thought of as the panacea to wildlife management challenges related to drought-related die-offs and the need for tourists to view the key wildlife species, is increasingly regarded as potentially highly detrimental to the resilience of semi-arid ecosystems unless provisioned and pumped strategically. In particular, AWPs can lead to the loss of migratory behaviour and the mobility of ungulates and their predators due to their tendency to stay around the water point (Owen-Smith, 1996; Sianga et al., 2017). This changes both the vegetation structure and composition, with profound changes to ecosystem functioning and detrimental impacts on rare species such as roan (*Hippotragus equinus*), sable (*Hippotragus niger*) and tsessebe (*Damaliscus lunatus*) and predators such as cheetah and wild dogs. They can also be dominated by elephants to the detriment of other species populations.

Based on respondents' views, waterholes are provided for tourism activities which put pressure on the tourism industry to provide water for their enjoyment. The potential conflict that AWPs create between the management of the Park's eco-system and that of its tourists should be emphasised, as most tourist operators relish the opportunity to view the big game that frequent AWPs all year-round. Critically, even with AWPs in place when drought hits, there will undoubtedly be large die-offs of wildlife, particularly elephants (Wato et al., 2016), simply due to the lack of forage.

In the case of fires in the CNP, respondents indicated that fires in the park eco-system are both a natural phenomenon, and an important phenomenon. Thus, the current issues with veld fires is that hot, late dry season fires tend to dominate (Cassidy et al., 2022) and are harmful to the environment, especially recruitment into the tree layer of some of the most valuable timber and fruit species. Burnt savannah is also not aesthetically pleasing to most tourists and so incompatible with photographic tourism.

One of the most critical issues in Protected Area management concerns the pro-vision of wildlife migratory corridors between wildlife landscapes so as to allow them to adapt to the changing climatic conditions (Perkins, 2019, 2020). Many spe-cies will have to shift their ranges as the climate warms, with dispersal into broader regions also greater able to accommodate seasonal fluctuations in forage availabil-ity. Unfortunately, many migratory corridors, such as some within the KAZA-TFCA are becoming closed off, or constrained by land use/land cover change as agricul-ture expands (Naidoo et al., 2018). It is an issue that is closely intertwined with the lack of benefits local people receive from living with wildlife at a time when Human-Wildlife Conflict (HWC) has reached unprecedented levels (Jones, 1999; Stoldt et al., 2020). Consequently, the short term benefits of subsidised agri-culture are all too often seen as supporting local livelihoods, while wildlife-based economies are seen as preventing economic development, except that of an invested elite (Perkins, 2020).

While there is widespread recognition amongst Policymakers that our climate is changing, few appear to grasp the scale and magnitude of the changes that are com-ing, and the implications it has for all sectors of the economy (Hambira et al., 2013; Perkins, 2020). As a result, most measures that are being implemented appear to operate at the wrong spatial scale (e.g. the current drive to boost game ranching in small fenced areas, rather than open unfenced rangelands) as well as misunder-stand the basics tenets of semi-arid ecology and the need for resilience (e.g. to facilitate migrations and ensure mobility of wild ungulates over large areas of connected landscape – rather than simply provide AWPs and pump them all the year round) (Perkins, 2021).

Past episodes of more arid conditions in Africa were adapted to by movements along what Balinsky (1962) termed the 'drought corridor' that connected the Kalahari-Namib region to eastern Africa via the Rift Valley and then with the Saharan-Sahelian zone (Perkins, 2020). It has led to calls for the establishment of today, what could be termed KALARIVA-TFCL, the Kalahari-Rift Valley Trans-Frontier Conservation Landscape that would connect Protected Areas between eastern and southern Africa, while also accommodating climate-smart agriculture and enabling multi-species based economies (Perkins, 2020). Unfortunately, without radical changes to the ways in which ecosystems are valued and conserved and for as long as local communities are largely excluded from the benefits of wildlife conservation, such transformative change will be impossible (Jones, 2002; Perkins, 2020).

12.5   Conclusions and Recommendations

Southern Africa is predicted to undergo considerable environmental changes in the near future. Global climate change is creating challenges for wildlife-based tourism and rural communities who are increasingly dependent on tourists (Saarinen et al., 2022). Based on the findings of this study, it is evident that environmental change is already causing negative impacts on wildlife-based tourism. Therefore, if not managed properly, it is likely to adversely affect the CNP, wildlife-based tourism and local employment by reducing the attractiveness of the place and, thus, the number of tourists. This is especially the case during the dry season, which has always been characterised by large numbers of wildlife – and also tourists. Thus, the results imply that environmental change in CNP is likely to affect the economic sustainability of the Park.

Ideally, sustainability of CNP depends on strengthening a wider socio-ecological resilience through the adaptation strategies such as the strategic provision and pumping of AWPs and maintenance and provision of migratory corridors; monitoring of drought and veld fires by the use of better technology and social innovations. Radio-tracking of elephants (Purdon et al., 2018), buffalo (Naidoo et al., 2014), and zebra (*Equus quagga*) (Bartlam-Brooks et al., 2011) have all revealed the importance of rainfall distribution as measured by EVI (Enhanced Vegetation Index), variations in which can trigger migrations over considerable distances, as well as more local movements within and between seasonal ranges. It follows that the provision and pumping of AWPs could take into consideration variations caused in EVI by rainfall and fires and so be managed so as to draw ungulates into areas where they can avoid drought-related mortality as well as high levels of HWC (Perkins, 2020). Blanket AWP provision and pumping all year-round, apart from sedentarising key ungulate and predator populations, runs the risk of holding animals in areas affected by drought, where they will inevitably succumb to starvation (Wato et al., 2016).

Migratory corridors at a large scale will also be essential for managing HWC and HEC within and around CNP, especially as southern Africa is projected to get drier and hotter. It remains that failure to look beyond the region in adapting to climate change may compromise the sustainability of the park environment, especially habitats, animals and the wildlife-based tourist experience. Thus, this study, therefore, recommends that there is a need to minimise environmental change impacts and unintended consequences by looking beyond local adaptation strategies and simply managing for the persistence of species at that scale. Further, there is a real need to ensure that CBNRM works and incentivises local communities to live with and tolerate wildlife due to the fact that it provides them with sustainable livelihoods and a chance to prosper (e.g. Lindsey et al., 2014). Currently, while a much-stated Policy goal and one that clearly provides the 'greatest good' (Child, 2002), CBNRM has not translated into meaningful benefits and inclusive development for local people with the inevitable result that agriculture is expanding and transforming large areas of rangeland, displacing wildlife and accentuating HWC. The need for integrated policies that lead to a sustainable balance between the wildlife and agricultural sectors, more equitable distribution of benefits and meaningful adaptation and resilience to climate change has never been greater.

References

Balinsky, B. I. (1962). Patterns of animal distribution on the African continent (summing-up talk). *Annals of the Cape Provincial Museum, 2,* 299–310.

Barnes, I. J. (2001). Economic returns and allocation of resources in the wildlife sector of Botswana. *South African Journal of Wildlife Research, 31*(3–4), 141–153.

Bartlam-Brooks, H. L. A., Bonyongo, M. C., & Harris, S. (2011). Will reconnecting ecosystems allow long-distance mammal migrations to resume? A case study of a zebra Equus burchelli migration in Botswana. *Oryx, 45*(2), 210–216.

Bio-Chobe. (2016). *Survey and assessment of the conservation threats of the Chobe National Park and Chobe Forest Reserves.* A report prepared for The United Nations Development Program.

BTO [Botswana Tourism Organization]. (2016). *The Chobe National Park.* Retrieved 19 September 2016 from: http://www.botswanatourism.co.bw/chobeNationalpark.php

Creswell, J. W. (2013). *Research design: Qualitative, quantitative, and mixed methods approaches.* Sage.

Campbell, J. M. (2008). Communicating for wildlife management or hunting tourism. In B. Lovelock (Ed.), *Tourism and the consumption of wildlife: Hunting, shooting and sport fishing* (pp. 213–226). Routledge.

Cassidy, L., Perkins, J. S., & Bradley, J. (2022). Too much, too late: Fires and reactive wildfire Management in Northern Botswana. *African Journal of Range & Forage Science, 39,* 160.

Child, B. (2002). The acceptable face of conservation. Wildlife conservation can bolster human needs rather than conflict with them. Book Review. *Nature, 415,* 581–582.

Child, B., & Barnes, G. (2010). The conceptual evolution and practice of community-based natural resource management in southern Africa: Past, present and future. *Environmental Conservation, 37*(3), 283–295. https://doi.org/10.1017/S0376892910000512

Chilembwe, J. M. (2020). Nature tourism, wildlife resources and community-based conservation: The case study of Malawi. In M. T. Stone, M. Lenao, & N. Moswete (Eds.), *Natural resources, tourism and community livelihoods in southern Africa: Chalenges of sustainable development* (pp. 26–38). Routledge.

Chiutsi, S., & Saarinen, J. (2017). Local participation in transfrontier tourism: Case of Sengwe community in Great Limpopo Transfrontier Conservation Area, Zimbabwe. *Development Southern Africa, 34*(3), 260–275. https://doi.org/10.1080/0376835X.2016.1259987

CSO (Central Statistics office). (2011). *Chobe sub-district population and housing census 2011: Selected indicators for villages and localities*. Statistics Botswana.

Cumming, D. H. M. (2008). Large scale conservation planning and priorities for the Kavango-Zambezi Transfrontier Conservation Area. In *A report prepared for Conservation International*. Conservation International.

Desanker, P., & Magadza, C. (2001). Africa. Climate change 2001: Impacts, adaptation and vulnerability. In J. J. McCarthy, O. F. Canziani, N. A. Leary, D. J. Dokken, & K. S. White (Eds.), *Climate change 2001: impacts, adaptation, and vulnerability*. Cambridge University Press.

Dillimono, H. D., & Dickinson, J. E. (2015). Travel, tourism, climate change, and behavioral change: Travelers' perspectives from a developing country, Nigeria. *Journal of Sustainable Tourism, 23*(3), 437–454.

Dobson, J. (2006). Sharks, wildlife tourism, and state regulation. *Tourism in Marine Environments, 3*(1), 15–23.

DWNP (Department of Wildife and National Parks). (2008). Chobe Management Plan, 2008. In *Ministry of environment, wildlife and tourism department of wildlife and national parks*. Government Press, Gaborone.

Earnshaw, A., & Emerton, L. (2000). The economics of wildlife tourism: Theory and reality for landholders in Africa. In H. H. T. Prins, J. G. Grootenhuis, & T. T. Dolan (Eds.), *Wildlife conservation by sustainable use* (pp. 315–334). Springer.

Fennell, D. (1999). *Ecotourism: An introduction*. Routledge.

Fox, A. D., Elmberg, J., Tombre, I., & Hessel, R. (2017). Agriculture and herbivorous waterfowl: A review of the scientific basis for improved management. *Biological Reviews*. https://doi.org/10.1111/brv.12258

GOB (Government of Botswana). (2001). *Botswana national atlas*. Department of Surveys and Mapping. Government Printer, Gaborone.

Engelbrecht, F., Adegoke, J., Bopape, M. J., Naidoo, M., Garland, R., Thatcher, M., & Gatebe, C. (2015). Projections of rapidly rising surface temperatures over Africa under low mitigation. *Environmental Research Letters, 10*(8), 085004. https://doi.org/10.1088/1748-9326/10/8/085004

Hambira, W., Manwa, H., Atlhopheng, J., & Saarinen, J. (2013). Perceptions of tourism operators towards adaptations to climate change in nature-based tourism: The quest for sustainable tourism in Botswana. *Botswana Journal of African studies, 27*(1), 69–85.

Hambira, W. L., Saarinen, J., & Moses, O. (2020). Climate change policy in a world of uncertainty: Changing environment, knowledge and tourism in Botswana. *African Geographical Review*. https://doi.org/10.1080/19376812.2020.1719366

Harper, A. R., Doerr, S. H., Santin, C., Froyd, C. A., & Sinnadurai, P. (2018). Prescribed fire and its impacts on ecosystem services in the United Kingdom. *Science of the Total Environment, 624*, 691–703.

Jones, T. B. (1999). *Community-based natural resource management in Botswana and Namibia: An inventory and preliminary analysis and progress*. Report submitted to International Institute for Environment and Development. IIED, London.

Jones, B. T. B. (2002). Chobe enclave: Lessons learnt from a community-based natural resources project 1993-2002. Ocassional paper N0.7, IUCN/SNV cbnrm support programme. IUCN, Gland.

Kilungu, H., Leemans, R., Munishi, P. K., Nicholls, S., & Amelung, B. (2019). Forty years of climate and land-cover change and its effects on tourism resources in Kilimanjaro National Park. *Tourism Planning & Development, 16*(2), 235–253.

Lepetu, J., & Garekae, H. (2020). Role of forest resources in local community livelihoods: Implications for conservation of Chobe Forest Reserve, Botswana. In M. T. Stone, M. Lenao, & N. Moswete (Eds.), *Natural resources, tourism and community livelihoods in southern Africa: Challenges of sustainable development* (pp. 176–189). Routledge.

Lindsey, P. A., Nyirenda, V. R., Barnes, J. I., Becker, M. S., & McRobb, R. (2014). Underperformance of African protected area networks and the case for new conservation models: Insights from Zambia. *PLoS One, 9*(5), e94109. https://doi.org/10.1371/journal.pone.0094109

Liu, Z. (2003). Sustainable tourism development: A critique. *Journal of Sustainable Tourism, 11*(6), 459–475.

Lovelock, B. (Ed.). (2008). *Tourism and the consumption of wildlife: Hunting, shooting and sport fishing*. Routledge.

Mabunda, D. M., & Wilson, D. (2009). Commercialisation of national parks: South Africa's Kruger National park as an example. In J. Saarinen, F. Becker, H. Manwa, & D. Wilson (Eds.), *Sustainable tourism in southern Africa: Local communities and natural resources in transition* (pp. 116–133). Channelview.

Mbaiwa, J. E. (2017). Poverty to riches: Who benefits from the booming tourism industry in Botswana. *Journal of Contemporary African Studies, 35*(1), 93–112. https://doi.org/10.108 0/02589001.2016.1270424

Mbaiwa, J. E. (2018). Effects of the safari hunting tourism ban on rural livelihoods and wildlife conservation in northern Botswana. *South African Geographical Journal, 100*(1), 41–61. https://doi.org/10.1080/03736245.2017.1299639

Mkiramweni, N. (2014). *Sustainable wildlife tourism in the context of climate change: The case study of Ngorongoro conservation area, Tanzania*. PhD Dissertation. Victoria University.

Mmopelwa, G., & Mackenzie, L. (2020). Economic assessment of tourism–based livelihoods for sustainable development: A case of handicrafts in southern and eastern Africa. In M. T. Stone, M. Lenao, & N. Moswete (Eds.), *Natural resources, tourism and community livelihoods in southern Africa: Challenges of sustainable development* (pp. 235–253). Routledge.

Mogende, E., & Moswete, N. (2018). Perceived wildlife-based tourism and impacts at the Chobe National Park, Botswana Wildlife. *PULA: Botswana Journal of African Studies, 32*(1), 48–67.

Mosugelo, D. K., Moe, S. R., Ringrose, S., & Nellemann, C. (2002). Vegetation changes during a 36-year period in northern Chobe National Park, Botswana. *African Journal of Ecology, 40*(3), 232–240.

Moswete, N., & Thapa, B. (2015). Factors that influence support for community-based ecotourism in the rural communities adjacent to the Kgalagadi Transfrontier Park, Botswana. *Journal of Ecotourism, 14*(2–3), 243–263.

Moswete, N., Nkape, K., & Tseme, M. (2017). Wildlife tourism safaris, vehicle decongestion routes and impact mitigation at the Chobe National Park, Botswana. In L. Ismar & R. Green (Eds.), *Wildlife tourism, environmental learning and ethical encounters* (pp. 71–88). Springer.

Moswete, N., & Thapa, B. (2018). Local communities, CBOs/trusts, and People–Park relationships: A case study of the Kgalagadi Transfrontier Park, Botswana. *The George Wright Forum, 35*(1), 96–108.

Naidoo, R., Du Preez, P., Stuart-Hill, G., Beytell, P., & Taylor, R. (2014). Long-range migrations and dispersals of African buffalo (*Syncerus caffer*) in the Kavango–Zambezi Transfrontier Conservation area. *African Journal of Ecology, 52*(4), 581–584. https://doi.org/10.1111/aje.12163

Naidoo, R., Kilian, J. W., Du Preez, P., Beytell, P., Aschenborn, O., Taylor, R. D., & Stuart-Hill, G. (2018). Evaluating the effectiveness of local-and regional-scale wildlife corridors using quantitative metrics of functional connectivity. *Biological Conservation, 217*, 96–103.

Newsome, D., Dowling, R. K., & Moore, S. A. (2005). *Wildlife tourism*. Channelview.

Novelli, M., Barnes, J. I., & Humavindu, M. (2006). The other side of the ecotourism coin: Consumptive tourism in southern Africa. *Journal of Ecotourism, 5*(1–2), 62–79. https://doi.org/10.1080/14724040608668447

Nyaupane, G. P., & Chhetri, N. (2009). Vulnerability to climate change of nature-based tourism in the Nepalese Himalayas. *Tourism Geographies, 11*(1), 95–119.

Nyirenda, V. R., Milimo, C., & Namukonde, N. (2020). Local people's perspectives on wildlife conservation, ecotourism and community livelihoods: A case study of Lusaka National park, Zambia. In M. T. Stone, M. Lenao, & N. Moswete (Eds.), *Natural resources, tourism and com-*

munity livelihoods in southern Africa: Challenges of sustainable development (pp. 108–122). Routledge.

Owen-Smith, N. (1996). Ecological guidelines for water points in extensive protected areas. *South African Journal of Wildlife Research, 26*(4), 107–112.

Perkins, J. S. (2019). 'Only connect': Restoring resilience in the Kalahari ecosystem. *Journal of Environmental Management, 249*, 109420. https://doi.org/10.1016/j.jenvman.2019.109420

Perkins, J. S. (2020). Take me to the river along the African drought corridor: Adapting to climate change. *Botswana Journal of Agriculture and Applied Sciences, 14*(1), 60–71. https://doi.org/10.37106/bojaas.2020.77

Perkins, J. S. (2021). Changing the scale and nature of artificial water point (AWP) use and adapting to climate change in the Kalahari of Southern Africa. In S. O. Keitumetse, L. Hens, & D. Norris (Eds.), *Sustainability in developing countries – Case studies from Botswana's journey towards 2030 agenda*. Springer Nature Switzerland AG.

Preston-Whyte, R., & Watson, H. (2005). Nature tourism and climatic change in southern Africa. In C. M. Hall & J. E. Higham (Eds.), *Tourism, recreation and climate change* (pp. 130–142). Challenview.

Purdon, A., Mole, M. A., Chase, M. J., & van Aarde, R. J. (2018). Partial migration in savannah elephant populations distributed across southern Africa. *Scientific Reports, 8*, 11331. https://doi.org/10.1038/s41598-018-29724-9

Reddy, M. V., & Wilkes, K. (2013). Tourism and sustainability: Transition to a green economy. In M. V. Reddy & K. Wilkes (Eds.), *Tourism, climate change and sustainability* (pp. 3–23). Routledge.

Saarinen, J. (2014). Critical sustainability: Setting the limits to growth and responsibility in tourism. *Sustainability, 6*(11), 1–17.

Saarinen, J., Fitchett, J., & Hoogendoorn, G. (2022). *Climate change and tourism in southern Africa*. Routledge.

Saarinen, J., Hambira, W. L., Atlhopheng, J., & Manwa, H. (2012). Tourism industry reaction to climate change in Kgalagadi South District, Botswana. *Development Southern Africa, 29*(2), 273–285.

Saarinen, J., Moswete, N., Atlhopheng, J., & Hambira, W. (2020). Changing socio-ecologies of Kalahari: Local perceptions towards environmental change and tourism in Kgalagadi, Botswana. *Development Southern Africa, 37*(5), 855–870. https://doi.org/10.1080/0376835X.2020.1809997

Santarém, F., Campos, J. C., Perreira, P., Hamidou, D., Saarinen, J., & Brito, J. C. (2018). Using multivariate statistics to assess ecotourism potential of water-bodies: A case-study in Mauritania. *Tourism Management, 67*, 34–46.

Santarém, F., Saarinen, J., Brito, J., & Pereira, P. (2019). New method to identify and map flagship fleets for promoting conservation and ecotourism. *Biological Conservation, 229*, 113–124. https://doi.org/10.1016/j.biocon.2018.10.017

Scott, D., Hall, C. M., & Gössling, S. (2012). *Tourism and climate change : Impacts, adaptation and mitigation*. Routledge.

Shoo, R., & Sorongwa, A. (2013). Contribution of ecotourism to nature conservation and improvement of livelihoods around Amani nature reserve, Tanzania. *Journal of Ecotourism, 12*(2), 75–89.

Sianga, K., van Telgen, M., Vrooman, J., Fynn, R. W., & van Langevelde, F. (2017). Spatial refuges buffer landscapes against homogenisation and degradation by large herbivore populations and facilitate vegetation heterogeneity. *Koedoe, 59*(2), 1–13.

Snyman, S., & Spenceley, A. (2012). Key sustainable tourism mechanisms for poverty reduction and local socio-economic development in Africa. *Africa Insight, 42*(2), 76–93.

Spenceley, A. (2008). Impacts of wildlife tourism on rural livelihoods in South Africa. In A. Spenceley (Ed.), *Responsible tourism: Critical issues for conservation and development* (pp. 159–186). Earthscan.

Stoldt, M., Göttert, T., Mann, C., & Zeller, U. (2020). Transfrontier conservation areas and human-wildlife conflict: The case of the Namibian component of the Kavango-Zambezi (KAZA) TFCA. *Scientific Reports, 10*, 7964. https://doi.org/10.1038/s41598-020-64537-9

UNWTO (World Tourism Organisation). (2008). *Policy for the growth and development of tourism in Botswana*. UNWTO/government of Botswana project for the formulation of a tourism policy for Botswana, July 2008. UNWTO and Department of Tourism, Gaborone.

van Wilgen, N. J., Goodall, V., Holness, S., Chown, S. L., & McGeoch, M. A. (2016). Rising temperatures and changing rainfall patterns in South Africa's national parks. *International Journal of Climatology, 36*(2), 706–721.

Wato, Y. A., Heitkönig, I. M. A., van Wieren, S. E., Wahungu, G., Prins, H. T., & van Langevelde, F. (2016). Prolonged drought results in starvation of African elephant (*Loxodonta africana*). *Biological Conservation, 203*, 89–96.

Wolf, A. (2009). *Preliminary assessment of the effect of high elephant density on ecosystem components (grass, trees, and large mammals) on the Chobe riverfront in Northern Botswana*. PhD Dissertation. University of Florida.

WTTC [World Travel and Tourism Council] (2015). *Travel and tourism economic impact 2015: Botswana*. London, UK. Retrieved from https://www.wttc.org/-/media/files/reports/economic%20impact%20research/countries%202015/botswana2015.pdf

WTTC. (2018). Travel and tourism economic impact study. www.wttc.org.

Maduo O. Mpolokang holds a MSc in Environmental Science from the University of Botswana with a keen interest in tourism development, climate change and environmental sustainability management. He is currently a Trainee Environmental Assessment Practitioner under Aqualogic Pty (Ltd), Botswana.

Jeremy S. Perkins is an Associate Professor in Range Ecology at the Department of Environmental Science at the University of Botswana. He undertook his PhD research on Kalahari cattleposts in 1988 and returned to Botswana in 1992. His research areas include the co-existence of wildlife and cattle, CBNRM, climate change and sand rivers.

Jarkko Saarinen is a Professor of Human Geography (Tourism Studies) at the University of Oulu, Finland, and Distinguished Visiting Professor (Sustainability Management) at the University of Johannesburg, South Africa, and Extraordinary Professor at the Tourism Management Division, Department of Marketing Management, University of Pretoria. His research interests include sustainable development, sustainable tourism, tourism-community relations and nature conservation studies.

Naomi N. Moswete is a Senior Lecturer in the Department of Environmental Science, University of Botswana. Her research interests include Human geography, tourism as a strategy for rural development, community-based tourism; Transboundary conservation areas –ecotourism nexus, parks–people relationships, heritage management & cultural tourism.

Chapter 13
The Impact of Rhino Poaching on the Economic Dimension of Sustainable Development in Wildlife Tourism

Berendien Lubbe

13.1 Introduction

Sustainable tourism is essential for conservation in protected areas such as the Kruger National Park in South Africa. The poaching of animals, particularly those regarded as iconic attractions for tourists, and the anti-poaching measures taken, impact tourism in the immediate future and the long term. However, the nature and depth of these impacts are still largely unknown. A question that is often raised and captures the complexity of the interaction between endangered wildlife and tourism is whether tourism would increase in the short term because tourists want to "see the last of the species" or decrease due to less opportunity for sightings? If tourism decreases over the longer term, this means less funding for conservation and socio-economic development in communities. Conservation agencies have recognized that endangered wildlife should be protected for their intrinsic and ecological value and ability to draw tourists generating much-needed conservation funds.

This chapter looks at the impacts and implications of rhino poaching on sustainable wildlife tourism in national parks in South Africa. The chapter begins with an overview of wildlife tourism and its economic consequences followed by discussing tourists' willingness to pay for wildlife experiences. The concept of wildlife poaching is described, followed by an overview of how rhino poaching has become a priority issue in South Africa. The chapter also touches on matters relating to anti-poaching measures and the role of local communities.

B. Lubbe (✉)
Department Historical and Heritage Studies, University of Pretoria, Pretoria, South Africa
e-mail: berendien.lubbe@up.ac.za

J. Saarinen et al. (eds.), *Southern African Perspectives on Sustainable Tourism Management*, Geographies of Tourism and Global Change,
https://doi.org/10.1007/978-3-030-99435-8_13

13.2 Wildlife Tourism

There is a difference between nature tourists and wildlife tourists; the former focuses on the enjoyment of nature as a holistic feature and the latter on observing wildlife as primary motivation (Chan & Baum, 2007; Curtin, 2010; Reynolds & Braithwaite, 2001). Catlin et al. (2011) broadly view wildlife tourism as any tourist activity having wildlife as its primary focus of attraction. Activities can either be consumptive, such as hunting, or non-consumptive such as wildlife watching. Recreational non-consumptive wildlife activities can take either a captive or free-ranging form. Since large tracts of land are protected areas, specifically for wildlife conservation and management, this gives rise to non-consumptive, free-ranging wildlife activities. Duffus and Dearden (1990, p. 215) defined the non-consumptive free-ranging wildlife activities as "a human recreational engagement with wildlife wherein the focal organism is not purposefully removed or permanently affected by the engagement". Tisdell and Wilson (2001) say that non-consumptive wildlife-oriented recreational (NCWOR) tourism is a significant and popular segment of the tourism industry. Such tourism activities generate substantial economic benefits for the conservation of wildlife species.

Wildlife tourism draws substantial numbers of international and domestic tourists worldwide. In their study on what attracts people to wildlife tourism experiences, Ballantyne et al. (2011) suggest that the sensory and emotional nature of the wildlife experience and the desire to "reconnect with nature" is the primary motivation. Tapper (2006, p. 14) identifies the critical factors as "being able to experience animals in the wild, to observe their 'natural' behaviour, and to appreciate their beauty". More easily observed species, particularly larger ones and those that show dramatic behaviours, for example, predators, rare and exotic species and those that have become symbolic in some way, attract the public's attention (Skibins et al., 2012). Cong et al. (2014) confirm this by stating an increased demand for tourists to interact with unusual or endangered animals. Wildlife tourism presents opportunities to observe and interact with endangered, threatened or rare animals (Higham & Shelton, 2011; Orams, 2002). In South Africa, iconic animals, often called the 'Big Five', function as flagship species, forming the foundation for the wildlife tourism experience. There has been an increase in the demand for charismatic and accessible free-ranging animals. Wildlife tourists tend to have high degrees of knowledge about specific animals, spending extensive time specifically watching them and feeling a sense of achievement if they learn something new about the species (Curtin, 2010).

Wildlife tourists tend to possess a strong environmental ethic, focus on intrinsic motivations, and desire to show their dedication to the cause (Curtin, 2010). Different visitor categories are developed based on specific characteristics derived from these values. More specifically, these values include a primary interest in wildlife, a solid affection for individual animals; concern for the right and wrong treatment of animals; and concern for the value of animals (Kellert, 1980). Kellert further distinguishes between several types of wildlife tourists: the naturalistic, whose primary interest derives from an affection for wildlife and outdoors; the ecologists who are concerned for the environment; the humanistic who demonstrates a strong love for individual animals; the moralistic who focusses on the right and wrong treatment of animals; the scientistic, interested in the physical attributes and biological functioning of animals; the aesthetic who is interested in the artistic and symbolic characteristics of animals; the utilitarian who looks at the material value of animals or habitat; the dominionists who, typically in sporting situations, are interested in mastery and control of animals; and the negativistic who through their indifference, dislike or fear, avoids animals. Kellert says an individual may encompass more than one category. In addition, the desired levels and types of interaction with animals or simply the motivation to tick off the must-see animals also produce further subcategories of wildlife tourists (Chan & Baum, 2007; Reynolds & Braithwaite, 2001).

13.3 Economic Aspects of Wildlife Tourism

Catlin et al. (2013) say that environmental economics usually assesses natural assets, such as wildlife, within a framework of "Total Economic Value" (TEV). These include values related to the asset, i.e. direct use, indirect use, and non-use values. Direct use values relate to the economic benefit derived directly from the use of the asset. In the case of wildlife, the financial advantage accrues from capturing and selling the animal or from people viewing the wildlife who pay for the experience. Use values differentiate between consumptive use value (capture and sale) or non-consumptive use values (tourism revenues). In tourism wildlife watching, income can enter a country's economy at several points. Figure 13.1 depicts a simple model of the monetary flows associated with protected areas and tourism. It shows how tourist dollars enter the economy through payments made by tourists to tourism-related businesses, the protected areas they visit, and through taxes levied at the national or local level (Tapper, 2006, p. 25).

Wildlife tourism can generate income in several ways, which include payments made by the tourists, such as the entrance fees or donations for the guides, drivers and other staff who may accompany them, allocation of government revenues, as well as sales of services and products at the site (Tapper, 2006). Tisdell (2003, p. 86) states that tourists also pay for accommodation and other services to travel to the wildlife watching sites. Destinations often present tourists that visit a destination for wildlife watching with opportunities in other tourism activities or to see and experience additional aspects such as the country's heritage and culture. Tapper (2006) argues that tourists presented with these additional opportunities are encouraged to stay longer and spend more money in the country.

Wildlife tourism can provide direct financial support for nature conservation and local communities where it occurs. In South Africa, tourism contributes 80% of South African National Parks (SANParks) revenue. Its key mandate is conservation (SANParks, 2020). It also has the developmental support provided to neighbouring

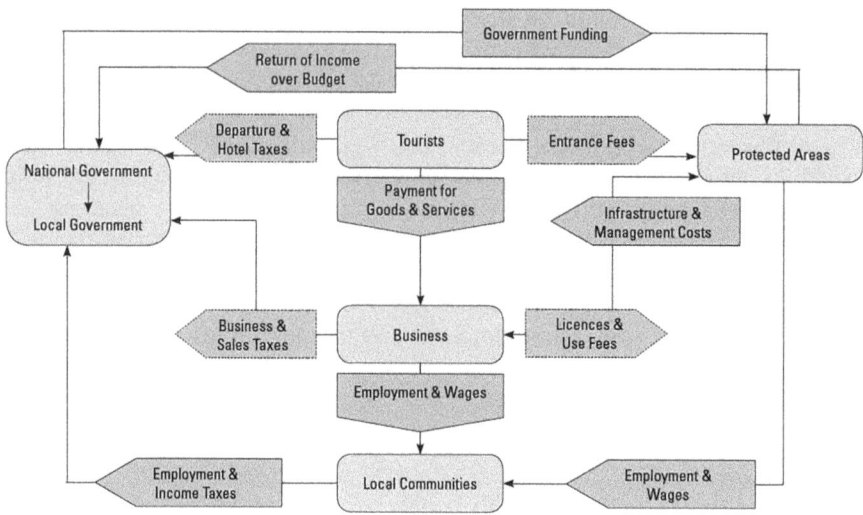

Fig. 13.1 A simplified model of the monetary flows associated with tourism and protected areas. (Modified from Tapper, 2006, p. 25)

communities as a priority. Many tourists find the presence of wildlife a significant reason for visiting a country or region or extending their stay, making wildlife watching a valuable asset for many destinations. The demand for wildlife tourism comes from a broad group of international and domestic visitors. Large numbers of people regularly pay significant amounts of money to view particular species of animals and nature. Tapper (2006, p. 24) estimated that 20% – 40% of all international tourists are interested in some form of wildlife watching. This interest ranges from enjoying casual observation of wildlife to taking short wildlife viewing excursions added to a trip undertaken primarily for other purposes, to tourists who spend their entire trips on wildlife watching. The UNWTO (2015) found that tourism to Africa has increased steadily, with an average annual growth rate of about 6.1% between 2005 and 2013. Tourist arrivals grew from 35 million in 2005 to reach a new record of 56 million in 2013. Numbers were predicted to more than double during the upcoming decade towards 134 million international arrivals in 2030.... provided, of course, they could still view elephants, rhinos, lions, gorillas and other iconic species.[1] More specifically, the global market size for wildlife tourism has been estimated at 12 million trips annually. It was growing at a rate of about 10% a year (UNWTO, 2015), with a wildlife watching tour typically comprising a group of six people and lasting approximately ten days. The iconic "Big Five" (African elephant, Cape buffalo, leopard, lion and rhino) are the main drawing cards in Southern and East Africa (Conservation Action Trust, 2015). Wildlife tourists are often willing to pay significantly more than current access fees for wildlife watching.

---

[1] As with the rest of the world's tourism arrivals predictions, this figure has been severely impacted by the consequences of the COVID-19 pandemic.

Booming wildlife tourism may also generate non-monetary benefits such as valuable political and government support for species conservation, support from local communities and key stakeholders, and public awareness of the significance of wildlife in the national heritage (Tapper, 2006).

13.4 Tourists' Willingness to Pay for Wildlife Viewing

According to Lipton et al. (1995), economic value measures the maximum amount an individual is willing to forego to obtain some good, service, or state of the world. An individual's willingness to pay (WTP) is the concept used to express this welfare measure. From a tourism perspective, willingness to pay is the maximum amount a person can pay to enjoy recreational facilities (McConnell, 1985). Several factors influence the willingness to pay regarding viewing wildlife, such as income, age, education, nationality, marital status, number of children, loyalty and donations (Saayman, 2014). For example, in Saayman's study, both education and income levels are positively related to willingness to pay. Still, education more powerfully than income or age may negatively or positively affect willingness to pay, depending on the purpose. Marital status generally shows a positive relationship with willingness to pay with married visitors willing to pay more than unmarried visitors. The study shows a negative correlation regarding nationality, with international visitors likely to pay more than local visitors (Saayman, 2014).

Tourism should be viable and profitable, meeting the standards expected by the market, including the design, price, reliability and quality of the services and experiences offered and its general attractiveness to competing products available at other sites. Management should understand the economic value derived from wildlife tourism and the different variables and factors influencing visitors' willingness to pay for viewing specific animals (like the 'Big 5'). This understanding will assist them in determining the visitor preferences, which will ultimately increase the tourism experience, achieving total tourist satisfaction. Providing visitors with quality experiences satisfying their needs will increase the demand for wildlife tourism, ensuring sustainability in the tourism and conservation sectors.

A study conducted in 2014 showed that wildlife was the most significant indicator of South Africa's unique product offerings (Lubbe et al., 2015), substantially more so than other product offerings such as culture and history, wine and food and even the welcoming nature and friendliness of the people. Wildlife experiences provide opportunities to observe and interact with animals. Being close to animals in their natural habitat fills tourists with fascination and wonder. The popularity and status of iconic animals, such as the Big Five, has a significant influence on the success of the wildlife experience for tourists. The tourist experience does not depend solely on the *actual* viewing of the animal but more so on the possibility that something fascinating may be waiting just around the corner (Tribe, 2009). Many parts of Africa rely heavily on its tourism businesses, with the most significant incentive for tourists being safaris that include the Big Five. Such tourism is currently under threat, with four of the five regarded as being either endangered or on their way to endangered status. Should one of the Big Five, namely the rhino, become extinct, followed by others, there may be far less tourism, resulting in fewer tour guides, drivers, lodge employees, restaurant employees, or souvenir shop employees. South Africa and Kenya are arguably two of the biggest benefactors of tourism via safaris, and with existing unemployment rates,[2] there is no room for lessening job opportunities (Wardlow, 2014). The World Tourism Organisation (UNWTO, 2015:7) states:

> ...wildlife crime is threatening the very existence of iconic species that are essential to Africa's image as home to the world's top wildlife destinations and thus jeopardizes the basis of one of Africa's most important tourism products. Security, safety, the conservation of ecosystems, and the quality of tourism products and services are basic prerequisites for successful tourism development, while poaching has serious negative impacts on the political, social and economic framework in which tourism development can take place. Consequently, the loss of wildlife caused by poaching is likely to significantly impact tourism development in Africa as well as the tourism sector worldwide linked to the African market with the subsequent reduction of the sustainable development opportunities linked to the sector.

13.5   Wildlife Poaching

> Poaching is the unlawful taking of wild animals or plants, whilst opposing domestic and international conservation and wildlife management regulations. Animals are usually killed for their hide, ivory, teeth, horns and bones which are sold to dealers to produce various products (Hall, 2012, p. 2).

As noted in the research on rhino poaching conducted for SANParks (Division Tourism Management Report, 2016), South Africa is one of the most biologically diverse nations globally. It has long promoted biodiversity conservation through the sustainable use of natural resources. South Africa's constitution enshrines these principles calling for: "*a prosperous, environmentally-conscious nation, whose people are in harmonious coexistence with the natural environment, and which derives lasting benefits from the conservation and sustainable use of its rich biological diversity*" (RSA Constitution, 1996). However, in South Africa, wildlife crime poses a significant threat to biodiversity, communities and tourism. It promotes ecological degradation, counteracts conservation efforts, and threatens the sustainable development and use of natural resources. It also exploits socio-economically vulnerable communities. Additionally, some communities on the borders of protected areas use socio-political issues to justify poaching as a form of protest (Gonçalves, 2017).

According to Hübschle (2017), the relationship between local people and parks in the South African context is complex due to historical, social and political factors. The study found that local people see conservation areas as symbols of elite

---

[2] In South Africa unemployment post-COVID stands at approximately 33% in 2021, according to Statistics South Africa.

interests and wealth, inaccessible to the poor majority. Despite efforts to garner community support and provide socio-economic development, those interviewed felt deprived of agency in co-determining projects and initiatives that would directly impact their lives and livelihoods.

Nearly three-quarters of the world's rhinos reside in South Africa (Johnson, 2014, p. 28), about 20,405 white rhinos and 5055 black rhinos. South Africa has the world's most successful conservation record for rhinos and in 2011 conserved 83% of Africa's rhinos and nearly three-quarters of all wild rhinos worldwide (Milliken & Shaw, 2012). Unfortunately, South Africa's strong conservation record of more than a century is threatened, with market forces in Vietnam, where the rhino population has become extinct, influencing the fate of South Africa's rhinos (Milliken & Shaw, 2012). Rhinos are poached for their horn, which can trade for $60,000 per kilogram and up to $180,000 on the black market (Carrington, 2014). Rising Chinese and Vietnamese demand for rhino horn have fuelled an upsurge in poaching. In Vietnam and China, rhino horn is falsely believed to be an anti-inflammatory in traditional 'medicine', and is a status symbol for the elite (Save the Rhino, 2021). Various strategies, campaigns, and anti-poaching measures arrested the escalation of rhino losses with a steady decline experienced from 2015 to 2019. Figure 13.2 provides an overview of the rate of rhinos poached in South Africa from 2007 to 2019.

According to the Minister of Environment, Forestry and Fisheries in South Africa, Ms Barbara Creecy, rhino poaching decreased by almost 53% in the first six months of 2020, a striking decrease compared to the same period in 2019 (Department of Environment, Forestry and Fisheries, 2020). This decline was due, in part, to the Covid-19 restrictions imposed during the first half of 2020.

As summarized in the research report on rhino poaching conducted by the Division Tourism Management (2016) on behalf of SANParks at the Kruger

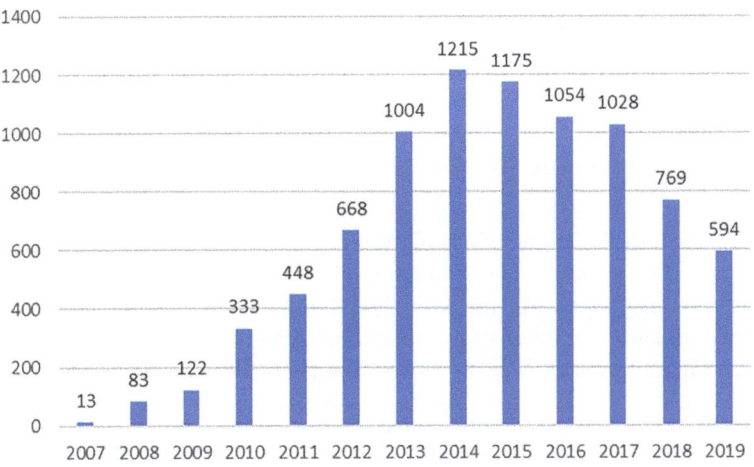

Fig. 13.2 Rhinos poached in South Africa 2007–2019. (Modified from Save the Rhino, 2021)

National Park and Hluhluwe-iMfolozi Game Reserve, the fight against rhino poaching is fiercest in and around the Kruger National Park. This Park covers 20,000 square kilometres of rough wilderness in the northwest corner of South Africa, with approximately half of the world's white rhinos located here. Other subspecies are scattered in small pockets of Asia and East Africa or private reserves, game farms and zoos (Johnson, 2014). The report (Division Tourism Management, 2016) notes that the poaching surge of the 1970s to the 1990s nearly completely wiped out the northern subspecies of the white rhinoceros except for a small remnant population numbering about 30 in the Democratic Republic of Congo's Garamba National Park on the border with Sudan. South Africa or Namibia housed most white rhino populations during this period (Milliken & Shaw, 2012). Vietnam was initially a rhino range state harbouring Asia's only surviving population of Critically Endangered Javan Rhinoceros (Rhinoceros sondaicus annamiticus). Still, by early 2010 the last animal had been poached for its horn (Milliken & Shaw, 2012). Vietnam's rhino horn trade has shifted to new sources in Africa. For nearly the past decade, South Africa has been the leading destination for resurgent illegal commerce out of Africa, especially from South Africa, where Vietnamese criminal operatives have become firmly embedded in the trade (Milliken & Shaw, 2012). The illegal wildlife trade is a big business. As De Rosner states, *"tourists come here and enjoy the beautiful bush and look at all the pretty animals, but what they don't know is that there is a full-blown insurgency going on here. This is a full-blown insurgency to save a species from extinction"* (in Johnson, 2014, p. 28).

The research report (Division Tourism Management, 2016) states that in the late 1950s, poaching and hunting had reduced South Africa's rhino population to 437 animals, resident in a limited spatial area of 72,000 acres that made them easy targets. Due to the efforts of conservationist Ian Player, the situation changed in the Umfolozi game reserve (now the Hluhluwe- iMfolozi) Park in Kwa-Zulu Natal Province. To control the numbers, he shipped some rhinos abroad, including to the USA, sent others to South Africa's game farms where they could safely mate all year. Against all odds, his scheme worked, and by the late 1960s, rhino numbers in South Africa had quadrupled to 1800.

13.6   The Economic Impact of Rhino Poaching and Tourism

While positive strides in the fight against rhino poaching are evident, the increase of poaching and rhino poaching continues to pressure South Africa's wildlife tourism market. As a result, it may be increasingly difficult to position itself as the preferred wildlife destination for international tourists. A report by the United Nations World Tourism Organisation (UNWTO, 2015) highlights the economic importance of wildlife tourism. It encourages tourism authorities and travel operators to fight against poaching and raise awareness of these issues among tourists. International tourism to Africa reached record levels in 2013, with 56 million tourists bringing in about R410bn. Significantly, 80% of them came to see the continent's wildlife. This valuable economic injection could increase by10% a year – provided poachers do not wipe out the iconic species that safari-goers travel to see. Threatened species survive mainly in protected areas; protected areas need money to remain operational, and tourism contributes to the conservation of these species in parks.

The report on rhino poaching (Division Tourism Management, 2016) concludes that the long-term effect of poaching on tourism is devastating from several perspectives: economically, socially and ecologically (UNWTO, 2015). Tourists associate Africa with the Big Five, and not being able to experience the Big Five would lead to a decline in tourists with the resulting economic implications of a decrease in profits, taxes and contribution to GDP. The tourism sector would experience a reduction of sustainable development opportunities linked to tourism. Employment opportunities for the local community involved in tourism sectors such as accommodation, restaurants and guiding would decline. The indirect benefits derived from redistributing protected area fees and community funding are at risk of creating more significant social problems. The increasing pressure on conservation efforts would result in higher prices for the wildlife experience but with a potential reduction in the element of value for money. The knock-on effects to the rest of the tourism supply chain, such as tour operators, would also be increasingly adverse. If not adequately addressed, the rising demand for exotic animals will move from one animal to another. If the rhino is extinct, the demand will shift to the next animal to fulfil the need.

Most studies on the economic impact of rhino poaching focus on the potential loss of tourism income and employment. Still, a question that arises in this debate is the monetary value or worth of the actual wildlife asset. In this respect, Catlin et al. (2013) say that attributing economic values derived from tourism industries to unique wildlife may be difficult because the fundamental characteristics of wildlife tourism activities do not lend themselves to valuation at the individual animal level. Allocating tourism values to animals through non-consumptive direct use-value can be highly speculative, unreliable, and lead to potentially misleading figures. There is not necessarily a direct and positive relationship between the presence of wildlife (or any particular species) and tourism value since according to Catlin et al. (2013):

- Tourists may spend the same amount on other goods and services regardless of the opportunity to see wildlife (or any particular species).
- The number of animals is not static. Changing the tourism experience can create the same value (e.g. if there are fewer animals, longer interaction time with fewer animals could deepen the experience, smaller numbers of animals would provide the greatest value per individual). Of course, if animals become extinct, tourists may not visit at all.
- Free-ranging species without definitive population estimates make up wildlife tourism; thus, the value of individual animals cannot be reliably estimated.
- The period over which an animal is valued for tourism can be misleading as tourists' views differ and reflect transient and returning animals.
- Wildlife tourism is focused on wildlife in general at the destination rather than the viewing of specific animals or species.

- The animal's behavioural responses to human presence can be misleading, with some animals avoiding proximity and will thus not be seen by tourists (behavioural responses could also change over time).
- Tourism values are particular to animals at an attraction and cannot be extrapolated to broader contexts.

Catlin et al. (2013, p. 97) further state that "*the very essence of valuing individual animals for the purpose of tourism is in itself a dubious process which appears to lend itself to misuse, possibly because of its potential to make good headlines.......* *The link between conservation of wildlife and tourism is strong and does not require embellishment*". The impact of rhino poaching on tourists themselves have specific short-term effects (Lubbe et al., 2015). While a spike in visitation may occur if visitors believe that rhinos will become extinct, there will be a decline in tourist visits if poaching and anti-poaching activities continue. Poaching and anti-poaching measures create a sense of unease and even fear as tourists experience the sight of poached animals and carcasses, hear gunshots, see more uniformed and armed rangers, hear overhead helicopters and experience car searches et cetera. Tourists sense of unease will directly affect tourism revenue. The research conducted by the Division of Tourism Management (2016) showed that the lack of effective communication by park management creates a forum for misinformation, which exacerbates the perception that poaching is out of control, leading to distrust and speculation by the public. Visitation and funding support is affected. Conservation education aims to enlighten the people on the importance of wildlife to the economy and conserve it. Programmes highlight the values and benefits attached to wildlife, why we should preserve it, and improve human-wildlife coexistence and tolerance (Kipng'etich, 2012).

Apart from the tourism-related loss of revenue due to rhino poaching, the costs of combatting such crimes are steep. Minnaar and Herbig (2018) list several expenses incurred to fight the rising number of poached rhinos, with the government, SANParks and private game farm owners forced to invest heavily in

- Additional armed game guards,
- surveillance and tracking equipment,
- improved security fencing and allied costs.
- The future sustainability of stocking parks and game farms with rhinos.

Other costs relate to decisions on the dehorning of adult rhinos, the development of rhino forensic investigation methods, and collecting evidence and DNA samples from all rhino poaching crime scenes by special units. Allied costs may be the deployment of the Defence Force to support the anti-poaching rangers.

The support of neighbouring communities in anti-poaching measures remains a contentious issue. Hübschle (2017, p. 440) says that "*in the eyes of the community, anti-poaching measures signify the social reproduction of historical inequalities, stigmatization and alienation of communities, who, under different circumstances and framing, might be agents of change and disruptors of illegal horn supplies*".

13.7   Concluding Remarks

While rhino poaching has seen a decline over the past few years, the long-term impacts of rhino poaching on tourism remain serious for several reasons. In the short term, a tourists chances of viewing rhinos (and other endangered species) diminishes since the animals are harder to find, and bad sightings may occur (e.g. poached carcasses of animals). Tourists begin to feel unsafe, resulting in a negative perception of the country. The negativity may result in fewer tourists reducing tourism receipts and thus affecting conservation efforts in the long term. From a social perspective, the lower tourist numbers affect communities by reducing employment and entrepreneurial activities, especially in developing countries where informal businesses within neighbouring protected areas are the drivers of economic activity.

References

Ballantyne, R., Packer, J., & Sutherland, L. A. (2011). Visitors' memories of wildlife tourism: Implications for the design of powerful interpretive experiences. *Tourism Management, 32*(4), 770–779.

Carrington, D. (2014, June). Fewer elephants killed in 2013, figures show. *Guardian*.

Catlin, J., Jones, R., & Jones, T. (2011). Revisiting Duffu's and Dearden's wildlife tourism framework. *Biological Conservation, 144*(5), 1537–1544. https://doi.org/10.1016/j.biocon.2011.01.021

Catlin, J., Hughes, M., Jones, T., Jones, R., & Campbell, R. (2013). Valuing individual animals through tourism: Science or speculation? *Biological Conservation, 157*, 93–98.

Chan, J. K. L., & Baum, T. (2007). Ecotourists' perceptions of ecotourism experience in lower Kinabatangan, Sabah, Malaysia. *Journal of Sustainable Tourism, 15*(5), 574–590.

Cong, L., Wu, B., Morrison, A. M., Shu, H., & Wang, M. (2014). Analysis of wildlife tourism experiences with endangered species: An exploratory study of encounters with giant pandas in Chengdu, China. *Tourism Management, 40*, 300–310.

Conservation Action Trust. (2015). News24, 25 June 2015.

Curtin, S. (2010). The self-presentation and self-development of serious wildlife tourists international. *Journal of Tourism Research, 12*, 17–33.

Department of Environment, Forestry and Fisheries. (2020, July 31). Retrieved from: https://www.environment.gov.za/mediarelease/rhinopoachingdecreases

Division Tourism Management. (2016, April 30). *The impact of rhino poaching on tourism*. Report on research conducted by the Division Tourism Management at the University of Pretoria, South Africa, on behalf of SANParks. South Africa.

Duffus, D. A., & Dearden, P. (1990). Non-consumptive wildlife-oriented recreation: A conceptual framework. *Biological Conservation, 53*, 213–231.

Gonçalves, D. (2017, June). Society and the rhino: A whole-of-society approach to wildlife crime in South Africa. *SA Crime Quarterly, 60*, 9–18.

Hall, C. M. S. (2012, October). *An investigation into the financial feasibility of intensive commercial white rhino farming in South Africa: A strategic approach*. Dissertation for the degree Bachelors of Industrial Engineering, University of Pretoria.

Higham, J. E. S., & Shelton, E. J. (2011). Tourism and wildlife habituation: Reduced population fitness or cessation of impact? *Tourism Management, 32*(6), 1290–1298.

Hübschle, A. M. (2017). The social economy of rhino poaching: Of economic freedom fighters, professional hunters and marginalized local people. *Current Sociology, 65*(3), 427–447.

Johnson, S. C. (2014, Jul/Aug). Where the wild things are. *Foreign Policy*, 28–32.

Kellert, S. (1980). American attitudes toward and knowledge of animals: An update. *International Journal for the Study of Animal Problems, 1*(2), 87–119.

Kipng'etich, J. (2012). Laying the foundation for conservation of Kenya's natural resources in the 21st century. *The GWS Journal of Parks, Protected Areas & Cultural Sites, 29*(1), 30–38.

Lipton, D. W., Wellman, K., Sheifer, I. C., & Weiher, R. F. (1995). Economic valuation of natural resources. *NOAA Coastal Ocean Program Decision Analysis Series No. 5*. [Online] Available from: http://www.csc.noaa.gov/coastal/economics/irreversibility.htm. Accessed: 2014-03-24.

Lubbe, B. A., Douglas, A., Fairer-Wessels, F., & Kruger, E. (2015, June). *A model to measure South Africa's tourism competitiveness*. Paper presented at the international Travel and Tourism Research Association (TTRA) Conference in Portland, Oregon, USA, 15–17.

McConnell, K. E. (1985). The economics of outdoor recreation. In A. V. Knesee & J. L. Sweeney (Eds.), *Handbook of natural resources and energy economics*. Elsevier Science.

Milliken, T., & Shaw, J. (2012). The South Africa – Viet Nam rhino horn trade nexus: A deadly combination of institutional lapses, corrupt wildlife industry professionals and Asian crime syndicates. Johannesburg: TRAFFIC.

Minnaar, A., & Herbig, F. (2018). The impact of conservation crime on the South African rural economy: A case study of rhino poaching. *Acta Criminologica: Southern African Journal of Criminology, 31*(4). https://journals.co.za/doi/abs/10.10520/EJC-159791abf8

Orams, M. B. (2002). Feeding wildlife as a tourism attraction: A review of issues and impacts. *Tourism Management, 23*(3), 281–293.

Republic of South Africa (RSA) Constitution. (1996). Available at: www.gov.za/documents/constitution/constitution-Republic-South-Africa-1996-1 (30 November 2015).

Reynolds, P. C., & Braithwaite, D. (2001). Towards a conceptual framework for wildlife tourism. *Tourism Management, 22*, 31–42.

Saayman, M. (2014). The non-consumptive value of selected marine species at Table Mountain National Park: An exploratory study. *South African Journal of Economic and Management Sciences, 17*(2), 184–193.

SANParks. (2020). *Annual report 2019–2020*. SANParks. Pretoria: South Africa.

Save the Rhino. (2021, April 17). Retrieved from: https://www.savetherhino.org/

Skibins, J. C., Hallo, J. C., Sharp, J. L., & Manning, R. E. (2012). Quantifying the role of viewing the Denali "Big 5" in visitor satisfaction and awareness: Conservation implications for flagship recognition and resource management. *Human Dimensions of Wildlife, 17*(2), 112–128.

Tapper, R. (2006). *Wildlife watching and tourism: A study on the benefits and risks of a fast-growing tourism activity and its impacts on species*. Retrieved from: http://www.cms.int

Tisdell, C. (2003). Economic aspects of ecotourism: Wildlife-based tourism and its contribution to nature. *Sri Lankan Journal of Agricultural Economics, 5*(1), 83–95.

Tisdell, C., & Wilson, C. (2001). Wildlife-based tourism and increased support for nature conservation financially and otherwise: Evidence from sea turtle ecotourism at Mon Repos. *Tourism Economics, 7*(3), 233–249.

Tribe, J. (2009). *Philosophical issues in tourism*. Channel View Publications.

UNWTO (World Tourism Organisation). (2015). Towards measuring the economic value of wildlife watching tourism in Africa. Briefing Paper. Madrid: UNWTO.

Wardlow, T. (2014). 50 million years on Earth… disappearing in 6?! Retrieved from: http://fightforrhinos.com.

Berendien Lubbe holds a PhD in Communication Management and is an Emeritus Professor and Research Associate in the Department of Historical and Heritage Studies at the University of Pretoria in South Africa. Her research currently focuses on contemporary issues in tourism. She has published in numerous journals, and her books on Tourism Distribution and Tourism Management in South Africa have been widely prescribed.

Chapter 14
Locational Heterogeneity in Climate Change Threats to Beach Tourism Destinations in South Africa

Jonathan Friedrich, Jannik Stahl, Gijsbert Hoogendoorn, and Jennifer M. Fitchett

14.1 Introduction

South Africa is the biggest tourism market in sub-Saharan Africa (United Nations World Tourism Organization [UNWTO], 2020) and is an essential tool for economic development (Rogerson, 2016). The coastline spans a range of climate zones (Lennard, 2019), ranging from subtropics to temperate conditions, each hosting various biomes (Finch & Meadows, 2019). The climate, fauna, and florae are vital attractions, primarily for coastal and beach tourism, which is highly dependent on the climate and natural setting and thus particularly vulnerable to climate change (Hoogendoorn & Fitchett, 2020).

The vulnerability to climate and environmental changes to tourism destinations indicates their threat level. This chapter does not aim to contribute to the debate around the concept of vulnerability specifically but instead use the applied concept of vulnerability to structure the chapter. Against this background, Moreno and

J. Friedrich
Leibniz Centre for Agricultural Landscape Research (ZALF), Müncheberg, Germany
e-mail: Jonathan.Friedrich@zalf.de

J. Stahl
Institute of Geography, University of Göttingen, Göttingen, Germany

G. Hoogendoorn (✉)
Department of Geography, Environmental Management and Energy Studies,
University of Johannesburg, Johannesburg, South Africa
e-mail: ghoogendoorn@uj.ac.za

J. M. Fitchett
School of Geography, Archaeology and Environmental Studies,
University of the Witwatersrand, Johannesburg, South Africa
e-mail: jennifer.fitchett@wits.ac.za

© The Author(s), under exclusive license to Springer Nature   199
Switzerland AG 2022
J. Saarinen et al. (eds.), *Southern African Perspectives on Sustainable Tourism Management*, Geographies of Tourism and Global Change,
https://doi.org/10.1007/978-3-030-99435-8_14

Becken (2009) defined a five-step methodology to assess the vulnerability of beach tourism destinations to climate change. This conceptual framework builds on the three-dimensional concept of vulnerability by evaluating the specific destination's exposure, sensitivity, and adaptive capacity (Agard & Schipper, 2015; Moreno & Becken, 2009). Vulnerability is unique to the geographical region under study in terms of *exposure, adaptive capacity,* and *sensitivity* (Füssel, 2007). This vulnerability leads to heterogeneous climate change threats to beach tourism destinations along the coastline of South Africa.

Moreno and Becken (2009) characterise *exposure* through the frequency and severity of environmental threats and the influence on social and material systems (Polsky et al., 2007). Polsky et al. (2007) define *sensitivity* through the perception and state of the social and material spheres. The *adaptive capacity* is structured through its available (economic) resources to develop sound strategies and policies to cope with current and future changes (Polsky et al., 2007). This notion of *adaptive capacity* has been criticised through the dynamic concept of social resilience, which proposes a triad of coping capacities, adaptive capacities, and transformative capacities (Keck & Sakdapolrak, 2013). This means it is focused on the material and immaterial aspects, supporting collectives, individuals or destinations in maintaining their social robustness to cope with current and (often uncertain) future threats and their different spatialities and temporalities (Keck & Sakdapolrak, 2013; Ziervogel et al., 2016). We structure this chapter around the applied concept of vulnerability that uses the term adaptive capacity in a non-dynamic manner. We follow this notion because a discussion of these concepts is not the aim of this chapter. At the same time, we acknowledge the benefits of focusing on social resilience and explicitly transformative capacity.

Exposure in terms of climate change stressors has been the field of study of many scholars in South Africa (e.g., Fitchett, 2018; Jung & Schindler, 2019; Jury, 2019; Pillay & Fitchett, 2019; Serdeczny et al., 2017). They have indicated that South Africa will experience severe climate change threats. These threats are heterogeneous and include, among others, changes in precipitation, increases in temperature, and an increased risk of droughts and floods. The *sensitivity* of beach tourists to climate change and the factors that describe the varying perceptions has been the subject of many studies in South Africa (e.g. Fitchett et al., 2016a, b; Fitchett & Hoogendoorn, 2018, 2019; Friedrich et al., 2020a, b; Giddy et al., 2017; Hoogendoorn et al., 2016). A complex set of factors determines the motivations and decisions of tourists, one of which is the destination's climate (Friedrich et al., 2020b; McKercher et al., 2015). Other factors include the personal situation of the tourist, including finances, timing, expectations (Rutty & Scott, 2013; Wilkins & de Urioste-Stone, 2018), and their country of origin (Friedrich et al., 2020a). The *adaptive capacity* of the South African tourism sector to climate change remains understudied. Still, it depends on factors such as the financial position of tourism operators and the extent to which the government perceives climate change as a risk (Hoogendoorn & Fitchett, 2018). Hoogendoorn and Fitchett (2018) argue that the *adaptive capacity* in the global South is lower than that of the global North, which further illustrates the increased risks climate change poses to the South African tourism sector. This aligns with Perch-Nielsen (2010) findings, who argue that South Africa's beach tourism is less vulnerable than small island states but more vulnerable than the global Norths beach tourism destinations. This not only relies on *sensitivity* and *exposure* but also on *adaptive capacity* in South Africa.

Spatial heterogeneity in climate change threats and perceptions is commonly known yet seldom explicitly explored in research. Against this backdrop, this chapter contributes to the growing body of literature exploring climate change impacts on tourism in South Africa, highlighting the heterogeneous climate change threats through analysing secondary data and findings of a quantitative questionnaire-based survey on climate perceptions of beach tourists. The research was conducted in November and December 2017 at a series of destinations along the South African coastline (cf. Friedrich et al., 2020a, b; see Figs. 14.1 and 14.2), namely Buffalo Bay, Durban, Cape St. Francis/St. Francis Bay, Cape Town, Jeffrey's Bay, Plettenberg Bay, Port Elizabeth, and St. Lucia, which cover the south (-east) coastline of South Africa. We highlight the urgent need to develop individual adaption strategies and policies for beach tourism destinations in South Africa to reduce the impacts of climate change, sustain the economic contribution, and foster a sustainable tourism sector.

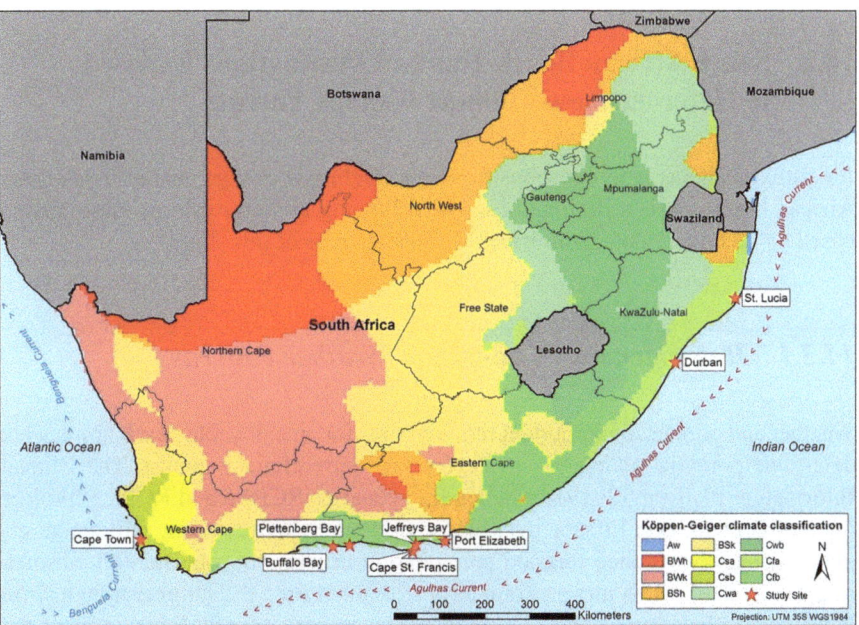

Fig. 14.1 South African climate zones (Köppen-Geiger), ocean-currents and study sites. (Adapted from Friedrich et al. (2020a), based on Peel et al. (2007))

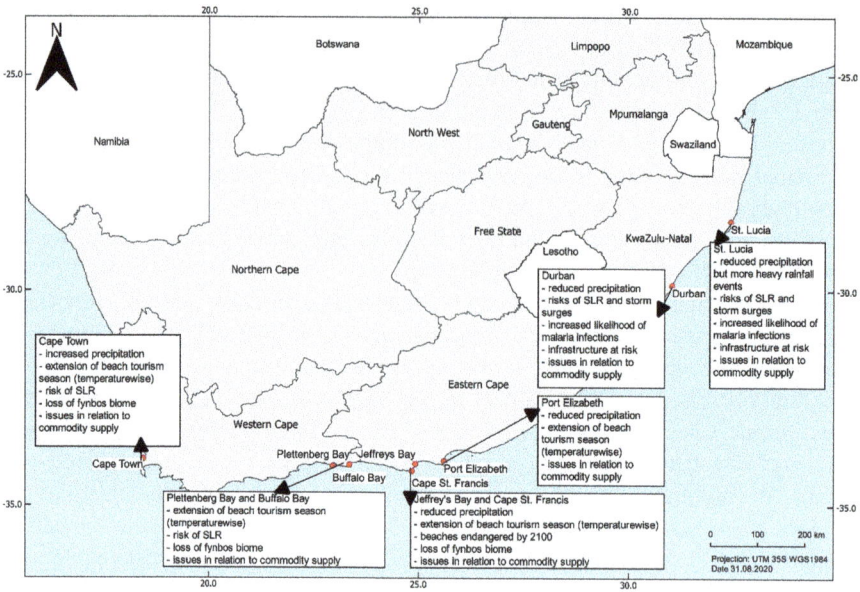

Fig. 14.2 Survey sites and summary of major anticipated climate change threats

14.2 South African Beach Tourism Destinations Exposed to Heterogeneous Climate Change Threats

As outlined above, heterogeneous *exposure* to climate change characterises South African beach tourism destinations. The following section explores these threats, based on secondary data, concerning the destinations of this study.

14.2.1 Heterogeneous Changes in Weather Patterns

Weather and climate are important requirements for an enjoyable beach holiday and determine potential adaptation strategies for future climate change. The Climate Information Platform (2019) modelled the scenarios RCP4.5 and RCP8.5[1] projecting that Durban, Port Elizabeth and Cape St. Francis/Jeffrey's Bay will, under climate change, experience reduced precipitation in terms of total rainfall, absolute rain days, rain days >5 mm and rain days >20 mm during the summer months from November to February in the 21st century. The scenarios for St. Lucia indicate that the destination is likely to experience less total precipitation and rain days but more heavy rainfall events, including rain days >5 mm and >20 mm. The projections for

---

[1] The Representative Concentration Pathways (RCPs) are scenarios based on change in radiative forcing in 2100 compared to the preindustrial time according to Taylor et al. (2012).

George (and, by proximity Plettenberg Bay/Buffalo Bay) indicate a decrease in precipitation, both in total rainfall and rain days. In contrast, heavy rainfall events are not projected to change. Cape Town is the only destination expected to experience an increase in precipitation. Here, an increase of heavy rainfall events is projected, such as rain days >5 mm and >20 mm among the months from November to February (CIP, 2019). However, these are notably outside of the rainfall season and reflect a minimal overall change in rainfall. Consequently, the Cape Town region is expected to experience an increased risk of droughts due to the changes in the position of the moisture corridors (Jury, 2019; Niang et al., 2015; Serdeczny et al., 2017), which the city and region already experienced in the 2015–2017 'Day Zero Drought'. Droughts can reduce the number of tourist arrivals long term due to both experiences of water restrictions and damage to the destination image (cf. Gössling et al., 2012; Smith & Fitchett, 2020). Droughts are not the only factor that may lead to a net decrease in tourist arrivals. The efficacy of water-saving policies and rules that apply in water scarcity, as found by Parks et al. (2019) for the drought of 2017–2018 in the Cape Town region, may also impact. However, a reduction in daily precipitation may lead to improved weather conditions for beach tourism initially, provided that the decline in precipitation does not heighten the occurrence of droughts (cf. Friedrich et al., 2020b).

Climate change projections (RCP4.5 and RCP8.5) for temperature show an increase in both daily maximum and minimum temperatures for all destinations, according to the CIP (2019). This temperature increase may not necessarily lead to a decline in tourist numbers in the case of South Africa. It could result in an extended beach tourism season at the destinations along the southwest coast as they will experience fewer days with temperatures below 22 °C (Friedrich et al., 2020b).

Temperature increases affect the thermal comfort of beach tourists while at the beach and increase the need for energy for cooling purposes, leading to additional costs for the accommodation sector (Roberts & O'Donoghue, 2013; Santos-Lacueva et al., 2017). These additional costs for the accommodation sector may vary between destinations along the coastline depending on the current temperature and existing facilities such as air conditioning.

Temperature and precipitation and wind, humidity, sunshine, and the interplay of these factors determine the comfort of Beach tourists. For example, Jung and Schindler (2019) project that South Africa will be one of the countries that experience increased mean wind speed, which could affect tourists' general comfort levels at the beach.

14.2.2 Heterogeneous Sea Level Rise Threatening Beach Tourism Destinations

Due to rising sea levels, the beach is threatened by erosion and inundation under climate change. Ocean currents exacerbate this threat (resulting in longshore drift) and the intensity and frequency of extreme events such as tropical cyclones and the associated storm surges (Fitchett, 2018; Pillay & Fitchett, 2019). Beach erosion affects the beach itself and accommodation establishments and infrastructure situated close to the beach (Colenbrander et al., 2015). Oppenheimer et al. (2019) project a global mean sea level rise of 0.84 m (RCP8.5) by the end of this century, while the regional sea-level rise for the South African coastline is estimated at 0.6 m (RCP4.5) and 0.7 m (RCP8.5) respectively during the same period (Carson et al., 2016).

Sea level rise will pose significant threats to many South African beach tourism destinations. Palmer et al. (2011) modelled that 41% of KwaZulu-Natal's swimming beaches, including St Lucia and Durban, are highly vulnerable in terms of their Coastal Vulnerability Index (CVI)[2] score. Research on sea-level rise in Cape St. Francis projects that the beaches will be in considerable danger by 2100 (Fitchett et al., 2016a, b). Musekiwa et al. (2015) used a GIS-based methodology to develop a coastal vulnerability map for South Africa which suggests that areas around Cape Town, Post St. Johns, and East London will be the most vulnerable in terms of sea-level rise. Durban and St. Lucia on the east-coast may face an increased risk of damage caused by the increasingly intense and poleward tracking South Indian Ocean tropical cyclones which could enhance local flooding and beach erosion (Fitchett, 2018; Fitchett & Grab, 2014; Pillay & Fitchett, 2019; Smith et al., 2007). The heightened risk of flooding poses dangers to infrastructure and accommodation establishments (Fitchett et al., 2016a, b). Shore protection acts as a response to sea level results in a loss in attractiveness of the beach that can have negative implications in decreasing tourist numbers (Fitchett et al., 2016a, b; Hamilton, 2007).

14.2.3 Heterogeneity in Indirect and Induced Climate Change Implications

In addition to the direct implications of climate change on beach tourism destinations in South Africa, the tourism sector may also experience indirect or induced effects. These are related to changes in vegetation, health, accessibility of destinations, and the production of commodities.

Climate change-induced shifts in precipitation and temperature patterns pose risks to ecosystems globally (Hall, 2018). In South Africa, climate change may result in shifts in multiple vegetation zones (Engelbrecht & Engelbrecht, 2016). The fynbos biome, an important touristic attraction located along the southern coast, including Cape Town, Buffalo Bay, Plettenberg Bay, Jeffreys Bay and Cape St. Francis, is considered the most vulnerable to climate change through shifts in precipitation and temperature patterns (Engelbrecht & Engelbrecht, 2016; Moncrieff et al., 2015).

---

[2] "Coastal Vulnerability Index (CVI) that divides portions of the coast into relative, predefined 'risk' classes." as defined by Palmer et al. (2011, p. 1390).

Tourism destinations face significant threats in terms of increased risks in health-related issues such as the south-westward expansion of the malaria region (e.g., Rosselló et al., 2017). This includes, but is not limited to, the northeast beach tourism destinations St. Lucia and Durban because of the increased likelihood of malaria infections (Ryan et al., 2015). Ziervogel et al. (2014) emphasise an increased risk of epidemics under climate change in South Africa. An example was the recent Listeria epidemic in 2018 – under climate change, future outbreaks of this disease is fostered due to challenges in food cold storage and transportation (Chersich et al., 2018). These epidemics can directly affect tourist health and may also encourage travel restrictions that can impact the accessibility of touristic destinations, as has been demonstrated during the COVID-19 pandemic (Rogerson & Rogerson, 2020).

The accessibility of destinations is stressed by climate change-induced storm surges that may destroy roads built close to the water (Smith et al., 2007). Storm surges may be the situation for St. Lucia and Durban, as they are under the highest risk of flooding, sea-level rise, and damage from the increased intensity of Mozambican tropical cyclones (Fitchett, 2018). In addition, climate change-induced heatwaves can affect the road infrastructure by melting the asphalt, which significantly increases the maintenance cost (Schweikert et al., 2015).

South African commodity production will face significant effects under climate change, impacting the demand for food products in the hospitality industry (Hoogendoorn & Fitchett, 2018; Rogerson, 2012). These threats include a heightened need for irrigation during droughts, increased pest risk, or negative impacts for plant growth under rising temperatures (Calzadilla et al., 2014; Phophi et al., 2020; Ziervogel et al., 2014).

14.3 Heterogeneous Sensitivity of the South African Beach Tourism Sector

The future development of both global and South African beach tourism sectors depends significantly on tourists' response to climatic changes (Scott et al., 2015). Against this background, this section analyses the heterogeneous climate perceptions of beach tourists. Friedrich et al. (2020a) and Friedrich et al. (2020b) present an in-depth analysis of the findings.

14.3.1 Methods

A questionnaire-based survey of 562 tourists collected the perceptions of climate and weather communicated by beach tourists staying at nine destinations (see Fig. 14.1). The survey took place for four weeks, from November 2017 to December 2017. It comprised responses from St. Lucia ($n = 100$), Durban ($n = 105$),

Port Elizabeth ($n = 51$), Cape St. Francis and Jeffrey's Bay ($n = 101$), Plettenberg Bay and Buffalo Bay ($n = 102$), and Cape Town ($n = 103$), covering the main beach tourism destinations along the eastern and southern coastline. According to data from municipalities, the remote western coast does not attract many tourists (Rogerson, 2017a). The nine destinations span a humid subtropical climate in the northeast to a warm summer temperate climate in Cape Town (Fig. 14.1). The destinations differ on the climate and the touristic attractions beyond the beach. St. Lucia attracts many tourists because of the significant nearby attraction of the iSimangaliso-Wetland-Park. Jeffreys Bay, Buffalo Bay, Cape St. Francis and Plettenberg Bay form part of the very popular Garden Route with various natural attractions such as waterfalls, bays, and nature reserves. With its mix of sea, beaches, and mountains, the internationally famous Cape Town stands in contrast to Durban, attracting a diversity of primarily domestic tourists who participate in various adventure tourism activities such as surfing, bungee jumping, and shark cage diving. In addition, every city and beach has its own colonial and apartheid history of segregation that continuously influences tourism and tourists' beach usage (Rogerson, 2017b).

The questionnaire consisted of quantitative questions on the importance of climate for various trip-related decisions and climatic and non-climatic factors potentially leading to trip cancellation, and climatic preferences related statements for the trip (for more information on the questionnaire, see Friedrich et al., 2020a). We used correlation analyses to prove the significant differences among the respondents' climate perceptions concerning the study site at which they have been staying. We used Cramer's V (nominal data; $p \leq \alpha \leq 0.05$) to quantify the results (de Lange & Nipper, 2018).

14.3.2 Heterogeneous Climate Perceptions at South African Beach Tourism Destinations

The respondents of this survey found the climate to be an important consideration in their travel decisions, while the extent of climatic influence differed between destinations. Generally, the weather was deemed most important when choosing South Africa as a destination and the trip's timing but was of least importance when selecting accommodation. Respondents vising Durban formed the largest cohort (36.2–47.6%; $n = 105$), indicating that climate was important when deciding whether to go on a trip. When summarising the categories 'very important' and 'important', respondents from Cape Town appear to be most concerned about the climate for decisions to go on a trip (e.g., 89.3% rated climate important or very important for travelling to South Africa). In contrast, respondents at St. Lucia and Port Elizabeth rated the climate least important for their decisions to go on a trip. The respondents from Cape St. Francis/Jeffrey's Bay and Plettenberg Bay/Buffalo Bay appear to be more concerned about the climate than respondents from Port

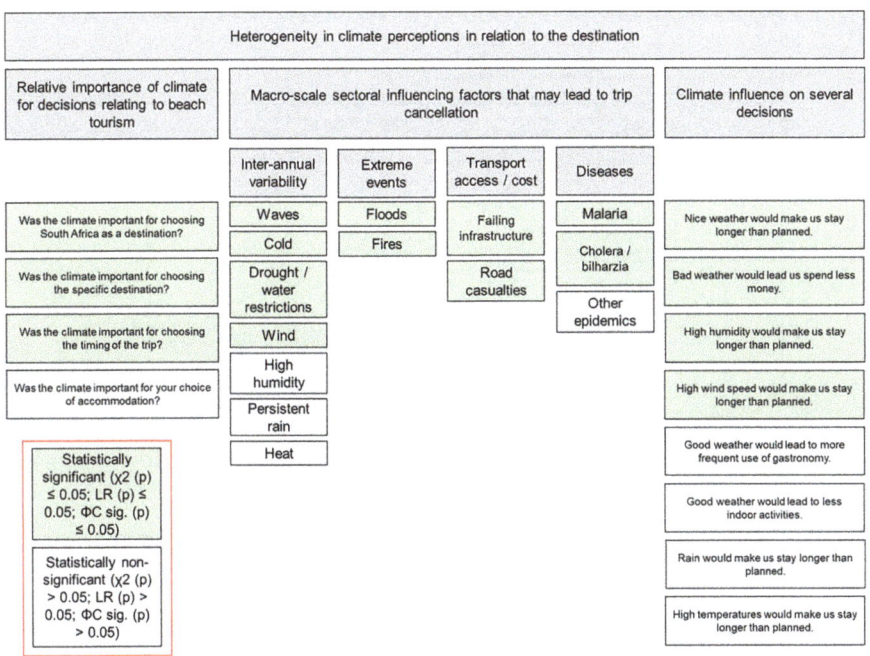

Fig. 14.3 Results of correlation analysis of climate importance for trip decisions; macro-scale sectoral influencing factors classified after Scott and Lemieux (2010); and climate-related statements concerning the destination of respondents; for values on statistical significance, see Friedrich and Stahl (2019)

Elizabeth and St. Lucia. However, they are less concerned than those respondents from Durban and Cape Town, two cities that may be under threat from climate change and increases in the prevalence of tropical cyclones and water scarcity, respectively (Fig. 14.3).

In terms of events that tourists considered to be potential trip cancellation factors, most respondents would not cancel in the case of climatic events related to inter-annual variability (such as droughts, water restrictions, or persistent rain) and transport access and cost-related issues (road casualties, failing infrastructure). Almost half of the respondents would not cancel their trip because of other potential threats, namely extreme events (floods, fires) and the outbreak of diseases (such as malaria and cholera). Figure 14.3 shows the heterogeneity in climate perceptions about the destination. It shows that perceptions of potential risk could lead to the cancellation of trips and that these perceptions differ according to destinations. For example, while a carefree stay is most important for respondents staying in Durban, a city experiencing several climate change threats, a high number indicated that they would cancel their trips in the case of extreme events, diseases, inter-annual variability, transport access and cost-related issues. The respondents visiting Port Elizabeth, a destination projected to experience fewer climate change threats than others, have been least concerned about climatic threats (inter-annual variability,

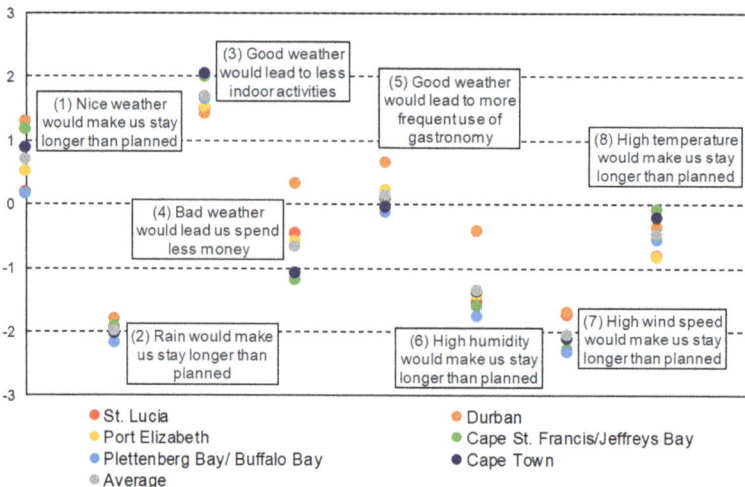

Fig. 14.4 Statement ranking from −3 (completely disagree) to +3 (completely agree) among destinations, ($n_1 = 512$; $n_2 = 510$; $n_3 = 545$; $n_4 = 498$; $n_5 = 477$; $n_6 = 496$; $n_7 = 530$; $n_8 = 526$)

extreme events) and transport access and cost. Although St. Lucia is considered a malaria region, most participants would not cancel their trip due to the occurrence of diseases, including malaria, suggesting that tourists visiting this destination already incorporate the risk in their decision-making.

Different climatic conditions (Fig. 14.4) affect tourists' travel plans. These lead participants of this survey to indicate that they would extend their stay under conditions of 'nice weather', do less indoor activities under 'good weather' conditions, and not extend their stay under rainy, humid, windy or high-temperature conditions. In contrast, there was greater variability in respondents' answers to the statement 'good weather would lead to more frequent use of gastronomy' (i.e. the activity of dining out). The same applies to the statement 'bad weather would lead us to spend less money'. For these cases, Durban's tourists indicate that they agree on these statements 4–6 (Fig. 14.4) more than respondents at other destinations. Correlation analyses confirm this heterogeneity among the destinations in correlation coefficients (Fig. 14.3).

The perceptions of climate and weather vary between destinations. As outlined in the introduction, a complex set of factors such as the motivation for travel, finances, the origin of tourists, and the risk perception paradox influence these differing perceptions (Friedrich et al., 2020a). Each of these factors may vary between destinations, and a range of tourists may visit each destination for primary attractions other than the beach. These varying factors add to the complexity of developing efficient adaptation strategies that address heterogeneity in both climate change threats and climate sensitivities of beach tourists. The complex vulnerability assessments methodology (Moreno & Becken, 2009) presents a solution through hazard-activity pairs. The methodological application in the South African context was recently tested (Friedrich et al., 2020b).

14.4 Conclusion

This chapter demonstrates that each beach tourism destination in this study will face multiple individual stressors in climate change. In addition, beach tourists perceive climate and weather heterogeneously depending on the destination they are staying. This heterogeneity underlines the importance of tourism operators and policymakers explicitly designing and implementing local and dynamic adaptation strategies. Climate change threats are heterogeneous, and the attractions and tourists' travel purposes may be affected under future climate change.

In general, extreme events and the increased likelihood of transmission of diseases under climate change may lead beach tourists to cancel their trips. In contrast, many tourists would not cancel their trip due to events considered to fall within inter-annual variability or issues of accessibility. Tourists visiting Durban appear to be most concerned about the climate and potential climate change threats. However, in contrast to the other destinations, these tourists seem to be least concerned about high humidity as a negative aspect for a trip extension since humidity already characterizes the current climate. This, in turn, means that unexpected weather events may change contemporary tourism habits at the specific destination – an aspect that needs reflection in future tourism planning.

Destinations are exposed to climate change heterogeneously. Durban and St. Lucia may especially be exposed to an increased risk of tropical cyclones, floods, sea-level rise, and heavy rainfall events, leading to high costs in road maintenance, destruction of touristic infrastructure, and potential decreases in tourism numbers. The southeast destinations are projected to experience heightened water stress and loss of local flora, leading to a loss in the region's attractiveness and potentially reducing tourist numbers. At the same time, warmer temperatures may extend this region's beach tourism season (Friedrich et al., 2020b). Climate change poses high adaptation costs for the tourism industry, possibly translating to higher prices of touristic services. Higher prices could, in turn, make tourism become an increasing issue of social distribution as only people with sufficient income may afford leisure trips. This may also negatively affect employees in the beach tourism sector if destinations do not have the economic or financial *adaptive capacity* to cope with climate change. Thus, adaption strategies need to focus on incorporating mitigation strategies such as sustainable tourism practices to foster a social-ecological transformation and align with the Sustainable Development Goals specified in the Agenda 2030. The concept of social resilience and explicitly transformative capacities prepares and transforms society and the tourism industry to cope with future threats in an inclusive and just sustainable way.

Acknowledgments Thanks to the CIP (2019) for offering free downscaled climate data for the whole African continent. The University of Johannesburg funded this study.

References

Agard, J., & Schipper, E. L. F. (2015). Glossary. In Intergovernmental panel on climate change (Ed.), *Climate change 2014: Impacts, adaptation and vulnerability, volume 2, regional aspects: Working group II contribution to the IPCC fifth assessment report* (pp. 1757–1776). Cambridge University Press.

Calzadilla, A., Zhu, T., Rehdanz, K., Tol, R. S. J., & Ringler, C. (2014). Climate change and agriculture: Impacts and adaptation options in South Africa. *Water Resources and Economics, 5*, 24–48. https://doi.org/10.1016/j.wre.2014.03.001

Carson, M., Köhl, A., Stammer, D., Slangen, A. B. A., Katsman, C. A., van de Wal, R. S. W., Church, J., & White, N. (2016). Coastal sea level changes, observed and projected during the 20th and 21st century. *Climatic Change, 134*(1–2), 269–281. https://doi.org/10.1007/s10584-015-1520-1

Chersich, M. F., Wright, C. Y., Venter, F., Rees, H., Scorgie, F., & Erasmus, B. (2018). Impacts of climate change on health and wellbeing in South Africa. *International Journal of Environmental Research and Public Health, 15*(9). https://doi.org/10.3390/ijerph15091884

Climate Information Platform. (2019). *African merged stations CMIP5*. http://cip.csag.uct.ac.za/webclient2/datasets/africa-merged-cmip5/

Colenbrander, D., Cartwright, A., & Taylor, A. (2015). Drawing a line in the sand: Managing coastal risks in the City of Cape Town. *South African Geographical Journal, 97*(1), 1–17. https://doi.org/10.1080/03736245.2014.924865

de Lange, N., & Nipper, J. (2018). *Quantitative methodik in der geographie. Grundriss allgemeine geographie: Vol. 4933*. Ferdinand Schöningh.

Engelbrecht, C. J., & Engelbrecht, F. A. (2016). Shifts in Köppen-Geiger climate zones over southern Africa in relation to key global temperature goals. *Theoretical and Applied Climatology, 123*(1–2), 247–261. https://doi.org/10.1007/s00704-014-1354-1

Finch, J. M., & Meadows, M. E. (2019). South African biomes and their changes over time. In J. Knight & C. M. Rogerson (Eds.), *The geography of South Africa* (pp. 57–69). Springer International Publishing.

Fitchett, J. M. (2018). Recent emergence of CAT5 tropical cyclones in the South Indian Ocean. *South African Journal of Science, 114*(11/12). https://doi.org/10.17159/sajs.2018/4426

Fitchett, J. M., & Grab, S. W. (2014). A 66-year tropical cyclone record for southeast Africa: Temporal trends in a global context. *International Journal of Climatology, 34*(13), 3604–3615. https://doi.org/10.1002/joc.3932

Fitchett, J. M., & Hoogendoorn, G. (2018). An analysis of factors affecting tourists' accounts of weather in South Africa. *International Journal of Biometeorology, 62*(12), 2161–2172. https://doi.org/10.1007/s00484-018-1617-0

Fitchett, J. M., & Hoogendoorn, G. (2019). Exploring the climate sensitivity of tourists to South Africa through TripAdvisor reviews. *South African Geographical Journal, 101*(1), 91–109. https://doi.org/10.1080/03736245.2018.1541022

Fitchett, J. M., Grant, B., & Hoogendoorn, G. (2016a). Climate change threats to two low-lying south African coastal towns: Risks and perceptions. *South African Journal of Science, 112*(5/6). https://doi.org/10.17159/sajs.2016/20150262

Fitchett, J. M., Hoogendoorn, G., & Swemmer, A. M. (2016b). Economic costs of the 2012 floods on tourism in the Mopani District Municipality, South Africa. *Transactions of the Royal Society of South Africa, 71*(2), 187–194. https://doi.org/10.1080/0035919X.2016.1167788

Friedrich, J., & Stahl, J. (2019). *Beach tourism and climate along South Africa's coastline: Climate awareness under threats of climate change and the socio-economic influence* [Unpublished Master Thesis]. Georg-August-Universität, Göttingen. https://doi.org/10.13140/RG.2.2.19690.98248

Friedrich, J., Stahl, J., Fitchett, J. M., & Hoogendoorn, G. (2020a). To beach or not to beach? Socio-economic factors influencing beach tourists' perceptions of climate and weather in South Africa. *Transactions of the Royal Society of South Africa, 75*(2), 194–202. https://doi.org/10.1080/0035919X.2020.1716869

Friedrich, J., Stahl, J., Hoogendoorn, G., & Fitchett, J. M. (2020b). Exploring climate change threats to beach tourism destinations: Application of the hazard-activity pairs methodology to South Africa. *Weather Climate and Society, 12*(3), 529–544. https://doi.org/10.1175/WCAS-D-19-0133.1

Füssel, H.-M. (2007). Vulnerability: A generally applicable conceptual framework for climate change research. *Global Environmental Change, 17*(2), 155–167. https://doi.org/10.1016/j.gloenvcha.2006.05.002

Giddy, J. K., Fitchett, J. M., & Hoogendoorn, G. (2017). Insight into American tourists' experiences with weather in South Africa. *Bulletin of Geography. Socio-Economic Series, 38*(38), 57–72. https://doi.org/10.1515/bog-2017-0034

Gössling, S., Peeters, P., Hall, C. M., Ceron, J.-P., Dubois, G., La Lehmann, V., & Scott, D. (2012). Tourism and water use: Supply, demand, and security. An international review. *Tourism Management, 33*(1), 1–15. https://doi.org/10.1016/j.tourman.2011.03.015

Hall, C. M. (2018). Climate change and its impacts on coastal tourism: Regional assessments, gaps and issues. In A. L. Jones & M. R. Phillips (Eds.), *Global climate change and coastal tourism: Recognising problems, managing solutions and future expectations* (pp. 48–61). CABI.

Hamilton, J. M. (2007). Coastal landscape and the hedonic price of accommodation. *Ecological Economics, 62*(3–4), 594–602. https://doi.org/10.1016/j.ecolecon.2006.08.001

Hoogendoorn, G., & Fitchett, J. M. (2018). Tourism and climate change: A review of threats and adaptation strategies for Africa. *Current Issues in Tourism, 21*(7), 742–759. https://doi.org/10.1080/13683500.2016.1188893

Hoogendoorn, G., & Fitchett, J. M. (2020). Fourteen years of tourism and climate change research in Southern Africa: Lessons on sustainability under conditions of global change. In M. T. Stone, M. Lenao, & N. Moswete (Eds.), *Natural resources, tourism and community livelihoods in southern Africa: Challenges of sustainable development* (pp. 78–89). Routledge. https://doi.org/10.4324/9780429289422-7

Hoogendoorn, G., Grant, B., & Fitchett, J. M. (2016). Disjunct perceptions? Climate change threats in two-low lying South African coastal towns. *Bulletin of Geography. Socio-Economic Series, 31*(31), 59–71. https://doi.org/10.1515/bog-2016-0005

Jung, C., & Schindler, D. (2019). Changing wind speed distributions under future global climate. *Energy Conversion and Management, 198*, 111841. https://doi.org/10.1016/j.enconman.2019.111841

Jury, M. R. (2019). South Africa's future climate: Trends and projections. In J. Knight & C. M. Rogerson (Eds.), *The geography of South Africa* (pp. 305–312). Springer International Publishing.

Keck, M., & Sakdapolrak, P. (2013). What is social resilience? Lessons learned and ways forward. *Erdkunde, 67*(1), 5–19. https://doi.org/10.3112/erdkunde.2013.01.02

Lennard, C. (2019). Multi-scale drivers of the South African weather and climate. In J. Knight & C. M. Rogerson (Eds.), *The geography of South Africa* (pp. 81–89). Springer International Publishing.

McKercher, B., Shoval, N., Park, E., & Kahani, A. (2015). The [limited] impact of weather on tourist behavior in an urban destination. *Journal of Travel Research, 54*(4), 442–455. https://doi.org/10.1177/0047287514522880

Moncrieff, G. R., Scheiter, S., Slingsby, J. A., & Higgins, S. I. (2015). Understanding global change impacts on south African biomes using dynamic vegetation models. *South African Journal of Botany, 101*, 16–23. https://doi.org/10.1016/j.sajb.2015.02.004

Moreno, A., & Becken, S. (2009). A climate change vulnerability assessment methodology for coastal tourism. *Journal of Sustainable Tourism, 17*(4), 473–488. https://doi.org/10.1080/09669580802651681

Musekiwa, C., Cawthra, H. C., Unterner, M., & van Zyl, F. W. (2015). An assessment of coastal vulnerability for the south African coast. *South African Journal of Geomatics, 4*(2), 123–137.

Niang, I., Ruppel, O. C., Abdrabo, M. A., Essel, A., Lennard, C., Padgham, J., & Urquhart, P. (2015). Africa. In Intergovernmental panel on climate change (Ed.), *Climate change 2014:*

Impacts, adaptation and vulnerability, volume 2, regional aspects: Working group II contribution to the IPCC fifth assessment report (pp. 1199–1265). Cambridge University Press.

Oppenheimer, M., Glavovic, B. C., Hinkel, J., van de Wal, R., Magnan, A. K., Abd-Elgawad, A., Cai, R., & Cifuentes-Jara, M. (2019). Sea level rise and implications for low-lying islands, coasts and communities. In H.-O. Pörtner, D.C. Roberts, V. Masson-Delmotte, P. Zhai, M. Tignor, E. Poloczanska, K. Mintenbeck (Ed.), *IPCC special report on the ocean and cryosphere in a changing climate* (pp. 321–445).

Palmer, B. J., van der Elst, R., Mackay, F., Mather, A. A., Smith, A. M., Bundy, S. C., Thackeray, Z., Leuci, R., & Parak, O. (2011). Preliminary coastal vulnerability assessment for KwaZulu-Natal, South Africa. *Journal of Coastal Research*, 1390–1395. http://www.jstor.org/stable/26482403

Parks, R., McLaren, M., Toumi, R., & Rivett, U. (2019). *Experiences and lessons in managing water from Cape Town*. Briefing paper No 29. https://www.imperial.ac.uk/grantham/publications/experiences-and-lessons-in-managing-water-from-cape-town.php

Peel, M. C., Finlayson, B. L., & McMahon, T. A. (2007). Updated world map of the Köppen-Geiger climate classification. *Hydrology and Earth System Sciences, 11*(5), 1633–1644. https://doi.org/10.5194/hess-11-1633-2007

Perch-Nielsen, S. L. (2010). The vulnerability of beach tourism to climate change—An index approach. *Climatic Change, 100*(3–4), 579–606. https://doi.org/10.1007/s10584-009-9692-1

Phophi, M. M., Mafongoya, P., & Lottering, S. (2020). Perceptions of climate change and drivers of insect pest outbreaks in vegetable crops in Limpopo Province of South Africa. *Climate, 8*(2), 27. https://doi.org/10.3390/cli8020027

Pillay, M. T., & Fitchett, J. M. (2019). Tropical cyclone landfalls south of the Tropic of Capricorn, southwest Indian Ocean. *Climate Research, 79*(1), 23–37. https://doi.org/10.3354/cr01575

Polsky, C., Neff, R., & Yarnal, B. (2007). Building comparable global change vulnerability assessments: The vulnerability scoping diagram. *Global Environmental Change, 17*(3–4), 472–485. https://doi.org/10.1016/j.gloenvcha.2007.01.005

Roberts, D., & O'Donoghue, S. (2013). Urban environmental challenges and climate change action in Durban, South Africa. *Environment and Urbanisation, 25*(2), 299–319. https://doi.org/10.1177/0956247813500904

Rogerson, C. M. (2012). Tourism–agriculture linkages in rural South Africa: Evidence from the accommodation sector. *Journal of Sustainable Tourism, 20*(3), 477–495. https://doi.org/10.1080/09669582.2011.617825

Rogerson, C. M. (2016). Climate change, tourism and local economic development in South Africa. *Local Economy, 31*(1–2), 322–331. https://doi.org/10.1177/0269094215624354

Rogerson, C. M. (2017a). Less visited tourism spaces in South Africa. *African Journal of Hospitality, Tourism and Leisure, 6*(3), 1–17.

Rogerson, J. M. (2017b). 'Kicking sand in the face of apartheid': Segregated beaches in South Africa. *Bulletin of Geography. Socio-Economic Series, 35*, 93–109. https://doi.org/10.1515/bog-2017-0007

Rogerson, C. M., & Rogerson, J. M. (2020). COVID-19 tourism impacts in South Africa: Government and industry response. *GeoJournal of Tourism and Geosites, 31*(3), 1083–1091. https://doi.org/10.30892/gtg.31321-544

Rosselló, J., Santana-Gallego, M., & Awan, W. (2017). Infectious disease risk and international tourism demand. *Health Policy and Planning, 32*(4), 538–548. https://doi.org/10.1093/heapol/czw177

Rutty, M., & Scott, D. (2013). Differential climate preferences of international beach tourists. *Climate Research, 57*(3), 259–269. https://doi.org/10.3354/cr01183

Ryan, S. J., McNally, A., Johnson, L. R., Mordecai, E. A., Ben-Horin, T., Paaijmans, K., & Lafferty, K. D. (2015). Mapping physiological suitability limits for malaria in Africa under climate change. *Vector Borne and Zoonotic Diseases (Larchmont, N.Y.), 15*(12), 718–725. https://doi.org/10.1089/vbz.2015.1822

Santos-Lacueva, R., Clavé, S. A., & Saladié, Ò. (2017). The vulnerability of coastal tourism destinations to climate change: The usefulness of policy analysis. *Sustainability, 9*(11), 2062. https://doi.org/10.3390/su9112062

Schweikert, A., Chinowsky, P., Kwiatkowski, K., Johnson, A., Shilling, E., Strzepek, K., & Strzepek, N. (2015). Road infrastructure and climate change: Impacts and adaptations for South Africa. *Journal of Infrastructure Systems, 21*(3), 4014046. https://doi.org/10.1061/(ASCE)IS.1943-555X.0000235

Scott, D., & Lemieux, C. (2010). Weather and climate information for tourism. *Procedia Environmental Sciences, 1*, 146–183. https://doi.org/10.1016/j.proenv.2010.09.011

Scott, D., Hall, C. M., & Gössling, S. (2015). A review of the IPCC fifth assessment and implications for tourism sector climate resilience and decarbonisation. *Journal of Sustainable Tourism, 29*(9), 1–23. https://doi.org/10.1080/09669582.2015.1062021

Serdeczny, O., Adams, S., Baarsch, F., Coumou, D., Robinson, A., Hare, W., Schaeffer, M., Perrette, M., & Reinhardt, J. (2017). Climate change impacts in Sub-Saharan Africa: From physical changes to their social repercussions. *Regional Environmental Change, 17*(6), 1585–1600. https://doi.org/10.1007/s10113-015-0910-2

Smith, T., & Fitchett, J. M. (2020). Drought challenges for nature tourism in the Sabi Sands Game Reserve in the eastern region of South Africa. *African Journal of Range and Forage Science, 37*(1), 107–117. https://doi.org/10.2989/10220119.2019.1700162

Smith, A. M., Guastella, L. A., Bundy, S. C., & Mather, A. A. (2007). Combined marine storm and Saros spring high tide erosion events along the KwaZulu-Natal coast in March 2007. *South African Journal of Science, 103*, 274–276. http://www.scielo.org.za/scielo.php?script=sci_artt ext&pid=S0038-23532007000400001&nrm=iso

Taylor, K. E., Stouffer, R. J., & Meehl, G. A. (2012). An overview of CMIP5 and the experiment design. *Bulletin of the American Meteorological Society, 93*(4), 485–498. https://doi.org/10.1175/BAMS-D-11-00094.1

United Nations World Tourism Organization. (2020). World tourism barometer and statistical annex. *UNWTO World Tourism Barometer, 18*(1). https://doi.org/10.18111/wtobarometereng

Wilkins, E. J., & de Urioste-Stone, S. (2018). Place attachment, recreational activities, and travel intent under changing climate conditions. *Journal of Sustainable Tourism, 26*(5), 798–811. https://doi.org/10.1080/09669582.2017.1417416

Ziervogel, G., New, M., van Archer Garderen, E., Midgley, G., Taylor, A., Hamann, R., Stuart-Hill, S., Myers, J., & Warburton, M. (2014). Climate change impacts and adaptation in South Africa. *Wiley Interdisciplinary Reviews: Climate Change, 5*(5), 605–620. https://doi.org/10.1002/wcc.295

Ziervogel, G., Cowen, A., & Ziniades, J. (2016). Moving from adaptive to transformative capacity: Building foundations for inclusive, thriving, and regenerative urban settlements. *Sustainability, 8*(9), 955. https://doi.org/10.3390/su8090955

Jonathan Friedrich is a Doctoral Researcher at Leibniz Centre for Agricultural Landscape Research (ZALF), Müncheberg, Germany. Research interests: social-ecological transformation, bioeconomy, environmental justice, sustainable tourism development and climate change.

Jannik Stahl is a Master of Science, Resource Analysis and Management, University of Göttingen. His research interests are tourism geography and climate research in South Africa.

Gijsbert Hoogendoorn is Professor in Human Geography at the Department of Geography, Environmental Management and Energy Studies, Johannesburg, South Africa. Research interests: climate change and tourism, second homes tourism, urban tourism.

Jennifer M. Fitchett is an Associate Professor of Physical Geography at the University of the Witwatersrand, Johannesburg. Her research is situated in the discipline of Biometeorology, exploring climate change and the impacts on plants, animals and people. This includes research into the effects of climate change on the southern African tourism sector.

Chapter 15
Sustainable Tourism Development Needs in the Southern African Context: Concluding Remarks

Jarkko Saarinen, Naomi N. Moswete, and Berendien Lubbe

15.1 Introduction

Sustainable tourism has become an established field of research in higher education across the southern African region. Hitherto, there is a growing body of regional literature on sustainable tourism with a substantial number of case studies that demonstrate both the success and challenges of tourism in the context of sustainable development. Therefore, this book is an excellent example of the diverse cases and issues in sustainable tourism development and management. In addition to research, many policies and strategies aim to support and encourage regional industries, local communities, and the public sector to introspect and collaborate towards more socially, economically, and environmentally sound tourism development. However, as Utting (2015, p. 1) noted, the core values and elements of the sustainability needs "often got lost in translation" when we are 'doing development' in practice.

Indeed, tourism development actions may too often emphasise short-term economic dimensions over social and environmental ones in practice (Sharpley, 2020). In contrast to this, sustainability needs should govern the industry and its related production and consumption circuit by guiding and regulating – when needed – the

J. Saarinen (✉)
Geography Research Unit, University of Oulu, Oulu, Finland

School of Tourism and Hospitality, University of Johannesburg, Johannesburg, South Africa
e-mail: jarkko.saarinen@oulu.fi

N. N. Moswete
Department of Environmental Science, University of Botswana, Gaborone, Botswana
e-mail: moatshen@ub.ac.bw

B. Lubbe
Department Historical and Heritage Studies, University of Pretoria, Pretoria, South Africa
e-mail: berendien.lubbe@up.ac.za

© The Author(s), under exclusive license to Springer Nature
Switzerland AG 2022
J. Saarinen et al. (eds.), *Southern African Perspectives on Sustainable Tourism Management*, Geographies of Tourism and Global Change,
https://doi.org/10.1007/978-3-030-99435-8_15

industry's growth and negative externalities (Reddy & Wilkes, 2013; Saarinen, 2018). This calls for the alternative ways to think about the relationships between tourism and localities, requiring governance and guiding frameworks for the industry that would support the implementation of sustainable innovations and activities in local and regional development (Bramwell & Lane, 2011; Bushell & Simmons, 2013; Saarinen & Gill, 2019). The SDGs may provide such a framework, though there is a need for more locally inclusive thinking and governance in development.

15.2   Tourism for Sustainable Development Goals

Based on this book's literature and case studies, many problems and emerging challenges exist in the tourism-sustainable development nexus. These challenges include the growing inequalities, poor inclusivity, evolving climate change impacts and related adaptation and mitigation needs, and the COVID-19 pandemic contesting the resilience of the industry, local communities and regional economies (Prideaux et al., 2020; Reddy & Wilkes, 2013; Rogerson & Baum, 2020). Despite the challenges, however, the past regional research and the case examples here demonstrate that it is possible to create synergies between tourism and localities in planning, development and management towards sustainable development. One of the critical frameworks for this could be the better implementation of tourism development to the SDGs (Saarinen, 2020; Scheyvens, 2018). However, as explicitly or implicitly indicated in the chapters of this book, the connections are still under-developed in the region (see also Dube & Nhamo, 2021; Siakwah et al., 2020).

Furthermore, there are challenging issues in implementing tourism to the high aims of the SDGs. According to Scheyvens et al. (2021, p. 3), for example, "there has been an insufficient focus to date on how the SDGs can best respond to local priorities and agendas in a tourism context or how different perspectives and voices represent the realisation of the goals." Indeed, tourism and sustainable development processes have a better chance of success when they recognise and support local needs (see Chiutsi & Saarinen, 2019; Manwa, 2003; Moswete et al., 2012; Rogerson & Rogerson, 2010). Thus, there is a call for more nuanced and sensitive approaches in sustainable tourism research and management. For example, instead of listing how many jobs tourism creates or could create in future (locally or globally), we should also analyse and be concerned about the kind of employment and for whom tourism creates these jobs (SDGs 8 and 10). We should be concerned about who is included in the development and increasing well-being of locally-based tourism (SDGs 5 and 10); and how global tourism development is locally implemented and driven (SDG 17) (Saarinen, 2020; UNWTO, 2017). These are relevant questions in any given destination context, but they are crucial, especially in the Global South. Tourism can provide economic growth in local communities, but to eradicate poverty (SGD1) there is a need to emphasise reducing local inequalities in development.

Therefore, to make positive contributions to the SDGs agenda, an all-inclusive and "well-designed and managed tourism sector" (UNWTO, 2017, p. 4) aligned to local and regional priorities is required. In this book, Godiraone et al., Maradza et al., Shereni et al., and Segobye et al., for example, highlighted the importance of understanding the local development context and needs. Simultaneously, however, there is also a need to develop a local understanding of the logic of the tourism industry, as indicated by Moswete et al., Nsanzya and Saarinen, and Steyn et al. Indeed, the book's core conclusion is the need to harness the relationship between tourism and the SDGs in a way that it considers respectful to local development needs, traditions and knowledge, as highlighted by Elijah et al., Green and Saarinen, and Harris and Botha.

15.3 Concluding Remarks: Towards Governing Resilient Tourism and SDGs Nexus for Localities

Current scholarship and the critical issues raised in the book emphasise the need to develop further the connections between sustainable development thinking and the tourism sector. SDGs offer a firm foundation for development, recognising inclusive forms of tourism that are more sensitive to local and regional power issues and relations in destination governance (see Pforr, 2004; Ramutsindela & Chauke, 2020). Based on the case examples in this book, there is a recommendation for more significant consideration of the local vulnerabilities, needs and priorities in tourism development. Furthermore, serious ongoing and emerging environmental and socioeconomic changes impact regional and tourism development. These changes will negatively affect the capacity of the tourism industry to contribute positively to the SDGs in southern Africa.

To support the tourism sector and SDGs nexus and decrease the industry's vulnerabilities, tourism-dependent communities and environment, their adaptive capacity elements need to be understood, researched and promoted. More specifically, the recent tourism and sustainable development literature have highlighted the idea of resilience (see Hall et al., 2018; Lew & Cheer, 2018; Ramutsindela & Mickler, 2020), which has become "one of the major conceptual tools [...] to deal with change" (Berkes & Ross, 2013, p. 6; see Tobin, 1999). Resilient tourism shows "a tourism system that has the adaptive capacity and resources to absorb change and disturbance and sustain itself by transforming via learning and innovation" (Saarinen & Gill, 2019, p. 24; see also Pforr, 2004; Tyrell & Johnston, 2008). It can help scientific communities and others understand the dynamics of complex, interwoven social, economic and ecological systems in southern Africa in the future (see Ramutsindela & Chauke, 2020; Siakwah et al., 2020). However, this calls for further research and development of suitable governance approaches for sustainable tourism that would be implementable and inclusive. In this way, the tourism industry would contribute positively to the SDGs and the African Union Commission's (2015) *Agenda 2063: The Africa We Want by* highlighting the need for inclusive growth, sustainable tourism, well-being and development in the near future.

References

African Union Commission. (2015). African Union Agenda 2063. The Africa we want. Final edition. https://au.int/en/agenda2063/overview

Berkes, F., & Ross, H. (2013). Community resilience: Toward an integrated approach. *Society and Natural Resources, 26*, 5–20.

Bramwell, B., & Lane, B. (2011). Critical research on the governance of tourism and sustainability. *Journal of Sustainable Tourism, 19*(4–5), 411–421.

Bushell, R., & Simmons, B. (2013). Facilitating sustainable innovations for SMEs in the tourism industry: Identifying factors of success and barriers to adoption in Australia. In M. V. Reddy & K. Wilkes (Eds.), *Tourism, climate change and sustainability* (pp. 42–57). Routledge.

Chiutsi, S., & Saarinen, J. (2019). The limits of inclusivity and sustainability in transfrontier peace parks: Case of Sengwe community in Great Limpopo Transfrontier Conservation Area, Zimbabwe. *Critical African Studies, 11*(3), 348–360. https://doi.org/10.1080/21681392.2019.1670703

Dube, K., & Nhamo, G. (2021). Sustainable development goals localisation in the tourism sector: lessons from Grootbos Private Nature Reserve, South Africa. *GeoJournal, 86*, 2191–2208. https://doi.org/10.1007/s10708-020-10182-8

Hall, C. M., Prayang, G., & Amore, A. (2018). Tourism and resilience. In *Individual, organisational and destination perspectives*. Channel View Publications.

Lew, A. A., & Cheer, J. (2018). *Tourism resilience and adaptation to environmental change: Definitions and frameworks*. Routledge.

Manwa, H. (2003). Wildlife-based tourism, ecology and sustainability: A tug-of-war among competing interests in Zimbabwe. *Journal of Tourism Studies, 14*(2), 45–54. https://search.informit.org/doi/10.3316/ielapa.200402853

Moswete, N., Thapa, B., & Child, B. (2012). Attitudes and opinions of local and national public sector stakeholders towards Kgalagadi Transfrontier Park, Botswana. *International Journal of Sustainable Development and World Ecology, 19*(1), 67–80. https://doi.org/10.1080/13504509.2011.592551

Pforr, C. (2004). Policy-making for sustainable tourism. In F. D. Pineda & C. A. Brebbia (Eds.), *Sustainable development* (pp. 83–94). WIT Press.

Prideaux, B., Thompson, M., & Pabel, A. (2020). Lessons from COVID-19 can prepare global tourism for the economic transformation needed to combat climate change. *Tourism Geographies, 22*(3), 667–678. https://doi.org/10.1080/14616688.2020.1762117

Ramutsindela, M., & Chauke, P. A. (2020). Biodiversity, wildlife and the land question in Africa: Sustainable Development Series. In M. Ramutsindela & D. Muckler (Eds.), *Africa and the SDGs*. Springer.

Ramutsindela, M., & Mickler, D. (2020). Global goals and African development. In M. Ramutsindela & D. Mickler (Eds.), *Africa and the SDGs* (pp. 1–9). Springer.

Reddy, M. V., & Wilkes, K. (2013). Tourism and sustainability: Transition to green economy. In M. V. Reddy & K. Wilkes (Eds.), *Tourism, climate change and sustainability* (pp. 3–23). Routledge.

Rogerson, C. M., & Baum, T. (2020). COVID-19 and African tourism research agendas. *Development Southern Africa, 37*(5), 727–741. https://doi.org/10.1080/0376835X.2020.1818551

Rogerson, C. M., & Rogerson, J. M. (2010). Local economic development in Africa: Global context and research directions. *Development Southern Africa, 27*(4), 465–480. https://doi.org/10.1080/0376835X.2010.508577

Saarinen, J. (2018). Beyond growth thinking: The need to revisit sustainable development in tourism. *Tourism Geographies, 20*(2), 337–340. https://doi.org/10.1080/14616688.2018.1434817

Saarinen, J. (2020). Tourism and sustainable development goals: Research on sustainable tourism geographies. In J. Saarinen (Ed.), *Tourism and sustainable development goals: Research on sustainable tourism geographies* (pp. 1–10). Routledge.

Saarinen, J., & Gill, A. M. (2019). Tourism, resilience and governance strategies in the transition towards sustainability. In J. Saarinen & A. M. Gill (Eds.), *Resilient destinations and tourism: Governance strategies in the transition towards sustainability in tourism* (pp. 15–33). Routledge.

Scheyvens, R. (2018). Linking tourism to the sustainable development goals: A geographical perspective. *Tourism Geographies, 20*(2), 341–342. https://doi.org/10.1080/14616688.2018.1434818

Scheyvens, R., Carr, A., Movono, A., Hughes, E., Higgins-Desbiolles, F., & Mika, J. P. (2021). Indigenous tourism and the sustainable development goals. *Annals of Tourism Research, 90*, 103260. https://doi.org/10.1016/j.annals.2021.103260

Sharpley, R. (2020). Tourism, sustainable development and the theoretical divide: 20 years on. *Journal of Sustainable Tourism.* https://doi.org/10.1080/09669582.2020.1779732

Siakwah, P., Musavengane, R., & Llewellyn, L. (2020). Tourism governance and attainment of the sustainable development goals in Africa. *Tourism Planning & Development, 17*(4), 355–383. https://doi.org/10.1080/21568316.2019.1600160

Tobin, G. A. (1999). Sustainability and community resilience: The holy grail of hazards planning? *Environmental Hazards, 1*, 13–25.

Tyrell, T. J., & Johnston, R. J. (2008). Tourism sustainability, resiliency and dynamics: Towards a more comprehensive perspective. *Tourism and Hospitality Research, 8*(1), 14–24.

UNWTO (World Tourism Organization). (2017). *Tourism and the sustainable development goals – Journey to 2030, highlights.* UNWTO.

Utting, P. (2015). Forewords. In P. Utting (Ed.), *Revisiting sustainable development* (pp. 1–15). UNRISD.

Jarkko Saarinen is a Professor of Human Geography (Tourism Studies) at the University of Oulu, Finland, and Distinguished Visiting Professor (Sustainability Management) at the University of Johannesburg, South Africa, and Extraordinary Professor at the Tourism Management Division, Department of Marketing Management, University of Pretoria. His research interests include sustainable development, sustainable tourism, tourism-community relations and nature conservation studies.

Dr. Naomi N. Moswete is a Senior Lecturer in the Department of Environmental Science, University of Botswana. Her research interests include Human geography, tourism as a strategy for rural development, community-based tourism; Transboundary conservation areas –ecotourism nexus, parks–people relationships, heritage management & cultural tourism. She is an editor in CABI Tourism Cases, sectional editor in Parks and Recreation Administration Journal.

Berendien Lubbe holds PhD in Communication Management and is an Emeritus Professor and Research Associate in the Department of Historical and Heritage Studies at the University of Pretoria in South Africa. Her research currently focuses on contemporary issues in tourism. She has published in numerous journals, and her books on Tourism Distribution and Tourism Management in South Africa have been widely prescribed.

Ingram Content Group UK Ltd.
Milton Keynes UK
UKHW020621210623
423802UK00002B/5